Edward Moore

The Time-References in the Divina Commedia and Their Bearing on

the Assumed Date and Duration of the Vision

Edward Moore

The Time-References in the Divina Commedia and Their Bearing on the Assumed Date and Duration of the Vision

ISBN/EAN: 9783337814106

Printed in Europe, USA, Canada, Australia, Japan

Cover: Foto ©Thomas Meinert / pixelio.de

More available books at **www.hansebooks.com**

THE TIME-REFERENCES

IN

THE DIVINA COMMEDIA

AND THEIR BEARING ON THE ASSUMED DATE

AND DURATION OF THE VISION

BY THE

REV. EDWARD MOORE, D.D.

PRINCIPAL OF S. EDMUND HALL, OXFORD

AND BARLOW LECTURER ON DANTE IN UNIVERSITY COLLEGE, LONDON

LONDON

DAVID NUTT, 270 STRAND, W.C.

1887

'The central man of all the world as representing in perfect balance the imaginative, moral and intellectual faculties all at their highest is Dante.'—J. RUSKIN.

'In all literary history there is no such figure as Dante.'— J. R. LOWELL.

Pontifical Institute of Mediaeval Studies

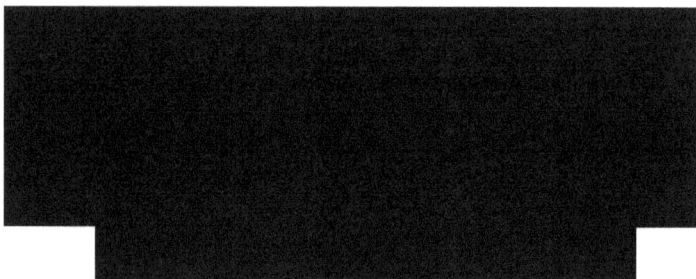

INTRODUCTORY NOTE.

THE substance of this Essay was delivered in two Inaugural Lectures in connexion with the Barlow Lectureship on Dante, in University College, London, in Nov. 1886.

I am not aware that the theory which I have here advocated has ever been applied consistently throughout the Poem before; and certainly not, I think, in any case with identity of results. The central principle itself has been occasionally recognized, and it is hardly necessary to say that in the interpretation of individual passages my conclusions or opinions have very frequently been anticipated. If it were not so, my work would at once stand condemned, considering the number and quality of previous writers on this subject:—οὐ γὰρ τούτους εὔλογον διαμαρτάνειν τοῖς ὅλοις, ἀλλ᾽ ἕν γέ τι ἢ καὶ τὰ πλεῖστα κατορθοῦν.

E. M.

S. EDMUND HALL, OXFORD,
    Dec. 15. 1886.

AUTHORITIES.

The principal authorities (as far as I am aware) on this subject are:—

GIAMBULLARI. Del Sito, Forma, e Misure dello Inferno di Dante (*Florence*, 1544).

DIONISI. Serie d' Aneddoti (*Verona*, 1785–90).

MAZZONI. Difesa della Commedia di Dante (*Cesena*, 1688).

PONTA. Nuovo Esperimento sulla Principale Allegoria della Divina Commedia (*Novi*, 1846).

CAPOCCI. Illustrazioni della Divina Commedia (*Naples*, 1856).

GRION. Che l' Anno della Visione di Dante è il MCCCI, &c. (*Udine*, 1865).

DELLA VALLE. Il senso Geografico-Astronomico dei luoghi della Divina Commedia (*Faenza*, 1869).

ANTONELLI. Studi Speciali (*Florence*, 1871).

PASQUINI. La Principale Allegoria della Divina Commedia (*Milan*, 1875).

PHILALETHES. Dante Alighieri's Göttliche Comödie (*Leipzig*, 1877).

LUBIN. Commedia di Dante Alighieri preceduta da Studi, &c. (*Padua*, 1881).

SCARTAZZINI. La Divina Commedia (*Leipzig*, 1874–82).

THE TIME-REFERENCES IN THE
DIVINA COMMEDIA

AND THEIR BEARING ON
THE ASSUMED DATE AND DURATION OF THE VISION.

THE references to details of time throughout the *Inferno* and *Purgatorio* are very numerous, and in many cases extremely obscure and difficult to interpret. At the same time, they are so pointed and definite in character that we are evidently intended to attach to them a precise meaning. Further, they clearly have relation to a comprehensive scheme or plan running throughout the whole poem; yet we find that besides the obscurity of individual passages, it is by no means easy (though I hope to show it is not impossible) to give a consistent and connected interpretation of these passages in relation to one another. They involve references sometimes to the position of the sun, but more frequently to that of the moon. The difficulties chiefly attach to the latter class, partly because the lunar movements are so much more varied and complicated, and still more because they of course entirely depend on the determination

of the day of the month and the year to which the
commencement of the Vision is to be referred.

As it has been disputed, at any rate by some recent
critics, whether that year is 1300 or 1301, and as
everything, in respect of references either to the
moon's position or to days associated with the Easter
Festival, will depend upon the year to which such
references belong, we cannot altogether avoid some
discussion upon this important initial point.

I wish however to make it clear at once, lest my
handling of this subject should be thought inadequate
(since many of the passages bearing on it do not involve 'time-references' in our present sense), that my
main purpose is to discuss the indications of the
hours by which the different stages of Dante's poetic
pilgrimage are marked, so as to fit them into a connected scheme; and more especially to deal with
those depending on lunar phenomena which involve
some difficulty in their interpretation. It is clearly
however impossible either to frame a connected
scheme, or to deal with the difficulties of isolated
passages, without a preliminary determination of
certain fundamental data on which the interpretation
of many of these passages in detail must depend.

Before then we attempt to bring together into a
general view (as I shall do later) the various time-
references to be found throughout the poem, we must
discuss, or at least endeavour to come to some under-
standing upon, these fundamental points, *viz.* the

year, the month, and the day, to be assumed for the commencement of the Vision.

One other brief explanation. Let it be remembered that I do not purpose to discuss astronomical references *as such*, but only when they convey data of time. Consequently many passages that are conspicuous in the elaborate works of Della Valle and others will find no place in the present Essay.

First of all then we must say something as to the *year*. The date of 1300 has been all but universally accepted from the time of the earliest Commentators down to the present day. The best-known advocate of the year 1301 is Grion, who, with much ingenuity and learning, maintained this, and one or two other paradoxes, in an elaborate monograph published in 1865.

It would be impossible for me (as I say) to undertake a full discussion of this point now, since Grion's theory depends on ingenious but questionable inferences drawn from a considerable number of passages, and chiefly such as involve prophecies (often very obscure and uncertain in their application) of events future to the assumed date of the Vision, *e.g.* notably *Inf.* vi. 67, x. 79[1], and others. Grion cannot cite

[1] Though I cannot discuss the question here, I may say that I believe the very difficult prophecy in *Inf.* x. 79 refers to the period of the departure of the Cardinal Niccolò da Prato from Florence on June 4, 1304, when the failure of the embassy of peace-making between the rival parties, on which he was sent by Benedict XI, was finally recognised, and the Bianchi were abandoned to the unrestrained fury of their enemies. If

any of the old Commentators as having pronounced for any other date than that of 1300. The most he can do is to claim the author of the *Ottimo Commento* (date 1334) on his side, not indeed explicitly, but by a hazardous inference from some of his presumably, because habitually, inaccurate statements. Boccaccio also, it is true, shows in one place some hesitation. In his Commentary on i. 1 [1] (in that interesting passage in which he relates Dante's account of his own age given to a friend on his death-bed) he declares distinctly for 1300; and so again in his explanation of vi. 69. But when commenting on iii. 60 (the celebrated 'gran rifiuto' passage) he excuses Dante for referring thus to Celestine on the ground that that Pope was not yet canonized; since Dante (continues Boccaccio) entered on this journey, as will appear in the twenty-first Canto of this Book, in 1301. Boccaccio is referring here to the important passage which we shall discuss presently, *Inf.* xxi. 112, but unfortunately his Commentary breaks off abruptly at *Inf.* xvii. 17. We cannot say therefore whether or not he has made a slip in this passing reference, or which of the two views to which he has inconsistently

so, this would imply the *popular* way of counting a lunation as equivalent to a *calendar* month (which is one of the disputed points in the interpretation of Dante's language in this passage; see Pasquini *Princ. Alleg.*, p. 239, &c.), and it would support *pro tanto* the adoption of a popular and not strictly scientific interpretation of other astronomical allusions, such as those discussed in the text later on.

[1] In Ed. Firenzi 1724, the references to the three passages are— pp. 19, 352, and 149 respectively.

committed himself he would ultimately have maintained, when brought to face the passage in question.
It will be noticed that in regard to the ground
alleged in iii. 60 (*viz.* the canonization of Celestine
being subsequent), the difference between the two
years is absolutely immaterial, since Celestine, who
died May 19th, 1296, was not canonized till May 5th,
1313.

Another advocate of the year 1301 is Vedovati
(*Exercitazioni Cronologiche*, &c., 1864), but the principal
reason for his adopting this view seems to be, that
he, being one of those who believe in a predominant
political signification of the poem, finds that 1301
(especially in reference to Canto I.) suits this theory
better than 1300 [1].

Per contra, among the latest and most distinguished
writers on Dante—Witte, Philalethes, Lubin, and
Scartazzini—all regard the date 1300 as quite beyond
question. The last-named says in one of his most
recent works, ' That the date of the Vision is the year
1300 is known to every one, nor should there arise on
this point the very slightest doubt.' To quote only
one other authority—the learned Dionisi, writing in
1788 (*Anedd.* iv. p. 45), says as to the year 1300,
' Niuno, ch' io sappia, ne dubita.'

This might perhaps suffice for our present purpose.
But I will briefly call attention to five or six passages
which convey irresistible conviction to my own mind;

---

[1] See Pasquini, *Princ. Alleg. della D. C.*, p. 231, and Suppl. Note
on Vedovati, *inf.* p. 131.

their plain and obvious meaning outweighing a host
of ingenious inferences based upon doubtful and
obscure prophecies, difficult astronomical allusions, or
questionable various readings. (1) We have the very
first line of the poem taken in connexion with the
well-known passage in *Conv.* iv. 23, where our human
life is compared to an arch, whose highest point is at
35 years; whence Christ willed to die in his 34th year,
so that His Divinity might not ' stare in descensione.'
(2) In *Inf.* x. 111 it is distinctly stated that Guido
Cavalcanti was still living. As he died during the
winter between 1300 and 1301, either late in 1300 or
early in 1301, this statement would be certainly untrue
in the Spring of 1301. (3) The language of Casella
in *Purg.* ii. 98–99 seems quite conclusive. He states
that the Indulgence connected with the Jubilee of
Boniface began just three months before, and that
during that time the spirits delayed outside the
entrance of Purgatory had felt the benefit of it.
This was actually proclaimed on Feb. 22nd, 1300,
but its privileges were antedated in the Bull itself
from the Christmas Day preceding. This clearly
necessitates the Spring of 1300, not 1301. (4) The
age of Can Grande della Scala is given in *Par.* xvii.
80 as ' pur nove anni,' ' only nine years.' Now as he
was born on March 9th, 1291, he would have been
just a little over ten years in the end of March or
beginning of April 1301 [1]. (5) In *Purg.* xxiii. 78

[1] Scartazzini's interesting note on this passage, *Par.* xvii. 80, should

Forese Donati says that five years have not yet passed since his death. He is generally said to have died in the end of 1295. It is true that Benvenuto da Imola (who however is distinctly admitted by Grion to have declared decisively for 1300 as the date of the Vision) gives 1296 as the date of Foresi's death. But it is a well-known source of confusion in these old dates (as I shall notice later on) that the year was sometimes held to commence on December 25th, sometimes on January 1st, and sometimes on March 25th, so that an event before March 25th, 1296, would be sometimes described as occurring at the end of 1295, and sometimes as at the beginning of 1296. (The so-called 'Revolution of 1688' in English History is a familiar illustration, as it occurred in February, 1689). (6) There is another, and I think very conclusive passage, which I do not see that Grion has noticed, in *Par.* ix. 40, when Dante says that the fame of his rather questionable Saint, Bishop Fulk (or Folchetto) of Marseilles, shall endure until

' Questo *centesim'* anno ancor s' incinqua.'

This expression more naturally applies to the initial

be read. Grion (p. 16), on the authority of one writer, maintains that Can Grande was born in May, 1280, and that by *nove anni* Dante refers to the years of Mars (in which planet he then was), which are rather less than double the length of ours. But apart from the inapplicability of the rest of the language of the passage if Can Grande were then about 21, nine revolutions of Mars will be insufficient, and of course more so for the date 1301 than for 1300. In another work (cited by Scart.) Grion proposes to read *dieci* for *nove*, but even this seems to be insufficient for his purpose.

year of a century (as it is popularly considered) than
to any other. (7) Finally, there is an argument not
depending on the interpretation of isolated passages,
but on a characteristic running like a thread through
the whole texture of the *Divina Commedia*, to which I
have not seen much weight given, *viz.* the well-known
fact that Dante never forgets the assumed date of his
Vision, and speaks of events which had then already
happened as past, but of all that had not yet
happened, even when they occurred very shortly
afterwards, as future, under the guise of prophecy.
It will scarcely be disputed (though it would take too
long to illustrate it in detail here) that that important
line of division between history and prophecy is
drawn at 1300 and not at 1301 ; nay, I believe we may
venture to say in April, or at least before May, 1300 [1].

Finally, there are many obvious reasons of a general
kind which suggest themselves for the selection of
this date. It was the central point of the ' arco ' of
Dante's own life (see *Conv.* iv. 23, already quoted): it
was the year of his Priorate at Florence, which was
the source of so much of his political troubles : it
was the beginning of a new century (at any rate
in the popular opinion and parlance): and it was also
the year of the first Jubilee of the Church [2].

We may now then I think confidently assume, as a

---

[1] Some examples in illustration of this argument will be found collected in a supplementary note.

[2] See further Dean Plumptre quoted *inf.* p. 114.

starting-point for the discussion of the *month* and *day* of the commencement of the Vision, that its *year* at any rate was, as is commonly supposed, 1300.

It will I think conduce to clearness, and also perhaps bespeak more interest in the discussion of the questions which follow, if I first set down briefly the chief landmarks which are clear and undisputed, to which therefore any solution must conform. I hardly dare to use the term 'undisputed' of any subject, or any passage, connected with this controversy, but I think there can be no reasonable dispute about the significance of the passages here collected, so far as I am proposing to employ them at present. Anyhow, disputed or undisputed, these are the data with which we are required to deal. They are of course drawn mainly from the *Inferno*, as we are dealing now with the question of the date of *commencement* of the Vision.

The central landmark, so to speak, is *Inf.* xxi. 112, from which it appears that it was then Easter Eve, it being universally agreed that the ruins, here and elsewhere referred to in the *Inferno*, resulted from the earthquake recorded at the moment of Christ's death. (This is in fact certain from *Inf.* xii. 34–45. See also *Par.* vii. 48 [1]). We shall find then that the night between Holy Thursday and Good Friday is supposed

---

[1] An interesting article will be found in Fornacieri's *Studi* (p. 31, &c.) on a theory of Benassuti, that the 'ruina' of *Inf.* v. 34 (a much disputed passage) is to be explained in reference to the same event.

to have been passed by Dante in the *Selva oscura:* see
i. 21—

> 'La notte ch' i' passai con tanto pieta.'

He emerges thence, and is encountered by the *Tre
Fiere* (whose significance has been so much disputed),
on the morning of Good Friday (see i. 37), the season
being that of Spring, and the sun among the same
stars as when he and they were first created (lines
38–40); *i. e.*, according to tradition, in the constellation Aries[1]. The whole day is spent in painful
hesitation and alternate advance and retreat, from the
dread of these three Beasts—(the long duration of
the conflict being indicated in lines 31–36, 59, &c.)—and
also in the interview with Virgil, who comes at last
to Dante's aid (l. 61, &c.)[2], so that it was nightfall on
Good Friday before they two together approached the
Entrance Gate of Hell (see ii. 1, &c., also lines 141–142,
and iii. 1, &c.).

Observe in passing how significantly Dante enters
the Inferno at nightfall, and both Purgatory and Paradise[3] at daybreak, and moreover the Earthly Paradise

---

[1] See further a supplementary note on this.

[2] Vellutetto expresses this very clearly in his note on *Purg.* xviii. 76 :
'Consumò il poeta tutto quel dì fin' alla sera in defendersi dalle fiere e
nel parlamento ch' ebbe con Virgilio.'

[3] See *Purg.* i. 19, &c.; *Par.* i. 43 ; and *Purg.* xxvii. 133, &c., respectively. It should be added that there is some dispute as to the last
point, Mr. Butler among others maintaining that Dante left the
Earthly Paradise, and entered the Heavenly Paradise *immediately*
after drinking the water of Eunoe, i. e. at *noon* on the Wednesday. I
still hold to the commonly received view as in the text. But even

as well. We find him leaving the fourth circle just after midnight (vii. 97–99), and passing from the sixth to the seventh circle between 3 and 5 a.m. on Easter Eve (see xi. 113, 114, compared with xv. 52 *ier mattina* &c.). He is leaving the fourth Bolgia of Malebolge (*i. e.* in the eighth circle) about sunrise, or (as he prefers to describe it) at *moonsetting* on Easter Eve (see xx. 125). He is in the fifth Bolgia of the same circle at 7 a.m., as appears from the very definite statement in xxi. 112 already referred to. It would be 7 or 10 a.m., according as the death of Christ is supposed to have taken place at the sixth or ninth hour, but that seems settled for us, as far as Dante is concerned, in favour of the sixth hour by *Conv.* iv. 23, as we shall see presently. We find him at the end of the ninth Bolgia of the eighth circle early in the afternoon of the same day, when the moon is directly under their feet (xxix. 10). He then traverses the ninth and last circle with its four divisions, and finally passes the centre of the earth to the other hemisphere, between 7 and 8 p.m. (see xxxiv. 68), which he suddenly finds now to be between 7 and 8 *a.m.* in that hemisphere; as is clearly indicated by the words in xxxiv. lines 96 and 105. We see then that the whole time occupied in traversing the Inferno is not much more than twenty-four or twenty-five hours, in fact from nightfall on the evening of Good Friday till a little after sunset on

if Mr. Butler be right, the symbolism to which I have called attention would be no less appropriate. See further note on p. 54.

Easter Eve. It should be noted however, that this
7–8 a.m. is not, as we might perhaps at first suppose,
the morning of Easter *Day*, but apparently the morning
of Easter *Eve* over again. This however is a disputed
point, which I shall discuss later on. Just, then, as
the whole of the daylight of Good Friday was passed
between the *Selva oscura* and the entrance of Hell,
so this intercalated space of twenty-four hours, or more
precisely of about twenty-one hours, is spent in passing from the centre of the earth to its surface at the
Mountain of Purgatory. When the poets emerged
'per un pertugio tondo,' it was 'riveder le stelle'
(last line of *Inf.* xxxiv.) ; *i.e.* the stars shining before
daybreak on Easter morning, as I shall maintain, or
on the morning of Easter Monday, as many others
hold. It was in fact about 5 a.m. in the morning ;
i. e. when Venus and the constellation of Pisces were
on the horizon, and the 'sweet hues of orient sapphire'
already in the sky, as we learn from the delicious
passage in *Purg.* i. 13–21. After the interview with
Cato, the sun is just rising (see ii. 1), and it is full and
brilliant day after the landing of the spirits from the
boat of the 'celestial nocchiero' (ii. 55): the beautiful
description of which scene is enhanced by the contrast
(as no doubt is intended in this and many other
parallel scenes and incidents in the *Inferno* and *Purgatorio*) with Charon, the 'nocchier della livida palude'
and his cargo of the 'mal seme d' Adamo' in *Inf.* iii.
98–117.

I will not, however, pursue this sketch of the poet's progress further into the Purgatorio at present. It will be seen that we have been able to follow his steps so far, almost from hour to hour by the help of passages which speak for themselves, *if* we can once determine our *terminus a quo*, *i.e.* the day of the ecclesiastical or civil year on which he assumes his journey to commence.

Now out of all these passages three points emerge clearly, and to these particular attention should be paid:—

(1) It was at the time of the Spring Equinox (i. 37–40) [1].

(2) He entered the Inferno the evening of the day after the Full Moon (xx. 127).

(3) The actual day was Good Friday (xxi. 112).

Here we have apparently three very clear and precise conditions, and so I believe we have really, if we merely take them all three in their simple, popular, and as I may say, natural sense. But unfortunately it is possible to understand every one of these apparently plain and precise data in two different senses. These may be roughly described as (1) the scientific or ideal sense, and (2) the popular

---

[1] This also appears from the curious and obscure passage in *Par.* i. 38, &c., where *quasi* in l. 44 seems intended to indicate (as Buti notes) that the Sun was not then exactly, but only approximately, at the entrance of Aries.

or natural sense. These I will now proceed to explain in each case.

I. As to the Spring Equinox.

This is, of course, generally and popularly understood to be March 21st. In this, as in other respects, the Calendar was regulated, chiefly with a view to the determination of Easter (since that is celebrated on the first Sunday after the first Full Moon after the Vernal Equinox)[1], by ecclesiastical authority, as early as the Council of Nicaea ; and by some 'particular and national' Churches at an even earlier date. As a starting-point, the Vernal Equinox is commonly believed to have been fixed (though this point is itself the subject of much controversy, into which we cannot enter here[2]), by that Council to be on

[1] It will be remembered that the Jewish months were strictly lunar, so that 'the fourteenth day of the first month,' the date prescribed for killing the passover (see Exodus xii. 2–6), would of course be the Full Moon of the first month. This explains the form of the present rule which took the place of the Quartodeciman practice. Easter was to be the first Sunday after the first Full Moon after the Vernal Equinox, at which time the year was thought to begin. Thus it was the best approximation that could be made to 'the fourteenth day of the first month,' allowing for (1) the fact that the months were no longer lunar, and (2) the condition that the day must fall on a Sunday. Hence among varying views as to the date of the Equinox—(March 18th, according to Hippolytus, c. 220 A.D. ; March 19th, according to Anatolius, c. 270 A.D. ; or March 21st, according to the view adopted either at or soon after the Council of Nicaea)—it was always insisted on that Easter should under no circumstances ever be kept *before* the Equinox, for thus it would no longer fall within 'the first month.' (See also an old Anglo-Saxon Chronicle quoted later on in the note on *Purg.* ii. 1–9, *inf.* p. 71.)

[2] This is not the place to discuss the thorny question as to the pre-

March 21st, though as matter of fact it occurred in the year 325, on the afternoon of March 20th, and though owing to the 'vibration' of the Equinox, consequent on the occurrence of Leap-years, and the inexact length of the Julian year, it might then fall actually on March 20th, 21st or 22nd[1]. According to the previous chronological arrangements of Julius Cæsar, it was on March 25th. Owing, however, to the large error in the Julian Calendar, which was afterwards corrected by the adoption of the New Style, in the year 1582, by the authority of Gregory XIII, the true Equinox in the time of Dante had fallen back as far as March 14th (or according to Dionisi, March 12), and this discrepancy between the true and the assumed Equinox was of course continually increasing, until by the time of Gregory's correction of the Calendar the Equinox had fallen back to March 11th, and consequently ten days had to be omitted from the Calendar to bring it back again to the 21st. This large and increasing error was

cise time and manner of the general adoption of the present practice of the Western Church for determining the observance of Easter. It seems to have originated with the Church of Alexandria, and to have differed from the practice of the Church of Rome, but by what precise means the latter church was persuaded to accept the rule of the former is much disputed. Those who wish to pursue this curious subject further will find ample materials in Hefele's *History of the Church Councils* (pp. 316-329 of Clarke's Translation, second edition), or in the elaborate *Appendix on the Paschal Controversy* in Dr. Butcher's learned work on *The Ecclesiastical Calendar.*

[1] See Butcher's *Ecclesiastical Calendar*, pp. 54-6.

certainly known to educated people in Dante's time. In 1267 Roger Bacon had calculated the Equinox as falling on March 13th, and had invoked the aid of Clement IV to correct the anomaly. (See Lubin's note, *Studi*, p. 362.) Moreover, Dante was himself perfectly aware of the error, since he alludes to it in *Par.* xxvii. 142, anticipating the time when January should pass entirely out of winter (*i.e.* when the Spring Equinox should fall back even beyond January into December) in consequence of the one-hundredth part of the day which is neglected on the earth,

> Per la centesma ch' è laggiù negletta,

the Julian year being too long by about that amount [1].

This then is the first question raised:—*Did Dante refer to the Equinox in its real and scientific, or in its ordinary and popular sense?* I shall maintain presently, *undoubtedly the latter*, but meanwhile let it be observed that this dispute exists as to the first of our data.

II. As to the Paschal Full Moon.

With a view to the determination of Easter, it was not only necessary to fix the Equinox, but also to adopt some 'mean' and not 'real' motions of the Moon, partly on account of the variability of the 'real' motions of the moon, and partly on account of differences of meridian. For this purpose approxi-

---

[1] See further supplementary note on this subject, p. 118.

mate Lunar Cycles have been adopted, the most celebrated of which is the Metonic Cycle to which the so-called Golden Numbers refer. Though it does not appear that this Cycle was formally recognized at Nicaea, yet it is found in use not long afterwards, and was possibly employed in the Alexandrian Church even in the third century. Dr. Butcher, in his learned and exhaustive work on the Ecclesiastical Calendar (to which I am indebted for many of these details), states that the Calendar Full Moon may differ as much as *two*, or sometimes even *three*, days from the real Full Moon. It is very important for our purpose to note that as a matter of fact it did so differ, by two days in the year 1300, and by even three days in 1301 [1].

It is to be observed then that the moon's position year by year was practically ascertained by the help of Cycles and rule-of-thumb calculations, such as those given in the Introduction to our Prayer Books, only of course much less accurate. There was no reference to, or verification by, independent astronomical observations.

Now the actual Full Moon by astronomical calculations for the year 1300 fell on Tuesday, April 5th, as is stated by Lubin, Scartazzini, Philalethes, and others. I may point to an obvious popular proof that this calculation is correct, from the fact that

---

[1] See the Calendars of this portion of the two years at the end of the Book.

there is an Eclipse of the Sun recorded as having
occurred on February 21st in that year ; and since

Solem quis dicere falsum audeat?

if we count the days from that date (remembering
that 1300 was a leap year), the next Full Moon but
one will be found to fall on April 5th. The same
result follows from working out the simple rule-of-
thumb calculations given by De Morgan in his *Book
of Almanacs*, p. xiii, for ascertaining the *real* New and
Full Moons in any given year.

On the other hand, it is no less certain that the
Full Moon *by the Calendar* fell upon Thursday April
7th, in the year 1300. This again may be found by
working out as before De Morgan's formulae for the
Calendar Moons from the Epact of the year, &c., or by
referring to any of the standard works of Chronology
in which the principal Epochs of past years are
given ; *e.g.* in the well-known *L'Art de vérifier les
Dates, Table Chronique*, the '*Terme Paschal*' (as it is
there called) for 1300 is placed on April 7th, and so
it will be found in any other similar works.

Here then is a second question, and one much
more vigorously disputed even than the last. *When
Dante speaks of the Full Moon, does he refer to the Real
(or Astronomical) or to the Calendar Moon?* And the
curious point about this is that some of the most
distinguished Commentators and Editors have (as we
shall presently see) missed the significance of this
discrepancy between the two Moons altogether, and

have assumed, *either* that Dante has made an unaccountable blunder as to the date of the Full Moon, *or* that for some reason or other his lunar references appear to be in strange and unexplained confusion.

III. As to the date of Good Friday and Easter.

According to a prevalent mediaeval belief, the actual day of the Crucifixion was March 25th, that event having taken place on the thirty-fourth anniversary of the Annunciation : *Eodem die* (says S. Cyril of Alexandria) *conceptus est in utero Christus et mortuus in Cruce*[1]. Moreover that day of the month was also thought to be the day of the Creation of the first Adam[2]. The inconvenience however of having a variable day of the *week* for Good Friday and Easter (even if the day could be ascertained with certainty), was thought to overbalance the propriety of having a fixed day of the *year*, and the varying of the day of the week was one serious objection to the Quartodeciman practice. From this point of view March 25th may be regarded as a sort of *ideal*, as opposed to the *conventional*, Good Friday. Again, the question

---

[1] This is quoted by Pasquini, *Princip. Alleg.*, p. 257. Dionisi also claims the authority of Tertullian, Lactantius, Augustine, Chrysostom, and others for the belief that the death of Christ took place on March 25th.

[2] Bede in his *Chron.* puts the creation of Adam (and also of Eve) and the Crucifixion, 'Eodem decimo Kal. Apr. (*i. e.* March 23rd). Decebat enim una eademque non solum hebdomadis sed et mensis die secundum Adam pro generis humani salute vivifica morte sopitum . . . qua videlicet die primum Adam . . . ipse creaverat, eique de latere costam tollens,' etc.

C 2

has been raised, and different answers have been given to it: *Did Dante adopt an ideal Good Friday, viz.* March 25, *or did he follow ordinary custom, and refer to Good Friday as generally observed, which would be in fact,* April 8 *in the year* 1300? One very precisely worded passage, *viz. Inf.* xxi. 112, has been thought to favour the former view. It is very far however from being decisive of the question, and regard must be had also to many other considerations [1].

There is yet another suggestion made under this head, and strange and improbable as it is, it is made, and supported (though with some hesitation), by no less an authority than Philalethes. He thinks it

---

[1] There is another mediaeval tradition assigning April 6th as the actual day of Christ's death, to which some interest attaches from its adoption by Petrarch. Moreover, though it does not seem to have made the basis of any theory making Dante's journey to commence on the night of April 5th in the *selva oscura* (it will be noticed that it is just one day wrong for Philalethes' suggestion, noticed afterwards in the text), yet its possible bearing on the question before us has not been wholly overlooked. (See supplementary note on Mazzoni's *Difesa, &c.*) Petrarch, with very questionable taste, dates the commencement of his affection for Laura from that solemn day. See *Son.* iii.—

> Era 'l giorno ch' al Sol si scoloraro
> Per la pietà del suo Fattore i rai
> Quand' i' fui preso, e non me ne guardai,
> Chè i be' vostr' occhi, Donna, mi legaro,

and again, *Son.* clvii.—

> Mille trecento ventisette appunto
> Su l' ora prima, il dì sesto d' Aprile,
> Nel Labirinto intrai; nè veggio ond' esca.

He refers to it also in some other passages. I should observe that April 6th was not the actual date of Good Friday in 1327, as it fell on April 10th in that year.

possible, from the difficulty (or supposed difficulty) attaching to the interpretation of one or two passages, that Dante may have followed a third method, *viz.* that by which the Jewish Passover was computed. This would have been sacrificed on the 5th of April in 1300, and the utterance of the words in *Inf.* xx. 127, *i.e.* 'the night before last the moon was full' would be on the 6th. Then the 'real' Full Moon being by calculation at 3 a.m. on the 5th, the expression *iernotte* as applied to it on the 6th in *Inf.* xx. 127 would be appropriately and naturally so applied: since though the Full Moon was astronomically on the 5th, yet occurring as it did at 3 a.m. it would be popularly described on the 6th as being on 'the night before last,' *i.e.* the night between the 4th and 5th.

Thus our three simple data (as they seemed to be) open a vista of much confusion, and we find at once the following different conceptions possible, and not only possible, but actually contended for, of our three, so to speak, fixed points. I will recapitulate them for the sake of clearness.

(1) *The Vision commenced at the time of the Spring Equinox.* If so, it may be held to have commenced either March 14th, or 21st, or 25th, or else at some time or other in the early spring, when the Sun was still in Aries, but not necessarily on the very day of his entering on that sign. Here are four different views, of which I shall maintain the last to be the correct one.

(2) *Dante entered the Inferno the day after the Full Moon at nightfall.* If so, it was the day after the *Real,* or the day after the *Calendar,* Full Moon; *i.e.* either on April 6th, or on April 8th. Certainly, as I hope to prove, it was the latter, *i.e.* April 8th.

(3) *He entered it on the Evening of Good Friday.* This again has been understood to signify either March 25th, April 5th, or April 8th. Again, as I think can be conclusively proved, it was on the last named, *viz.* April 8th; *i.e.* the actual Good Friday of the year 1300.

Now I have said each of these days (and I must add several others besides) have found their advocates. Critics have pounced upon one or two passages which seemed readily (or as they may have thought, crucially) to satisfy some one or more of these required conditions. They have then conveniently shut their eyes to other passages, and to the necessity of adopting a solution which should satisfy all the three conditions of the problem coincidently and simultaneously. Surely taken all together they form a threefold chain, which is not, or ought not to be, easily broken.

I must next point out what are the principal views, amidst all this variety, that have been held or seriously maintained, though I may be excused from examining any but a few of the most important or best supported of them. In fact, I may allege the plea and borrow the language of Aristotle (*Ethics,* Bk. I.

c. iv) : Ἄπασας μὲν οὖν ἐξετάζειν τὰς δόξας ματαιότερον ἴσως ἐστὶν, ἱκανὸν δὲ τὰς μάλιστα ἐπιπολάζουσας, ἢ δοκούσας ἔχειν τινὰ λόγον.

According to Dionisi (*Anedd.* iv. p. 45), Pierfrancesco Giambullari (who wrote a work on the *Inferno*, 1554) [1] was the first to ascertain by astronomical calculations that the Paschal Full Moon of 1300 was on *Monday* April the *fourth* (for so he puts it), and about *fifteen hours after midday*, i.e. what we should call 3 a.m. on *Tuesday* April the *fifth*; and he was therefore the first to find any difficulty or discrepancy in the data given by Dante in his poem. To meet this he adopts a theory that Dante intentionally disregards the facts with a view to some mystical propriety in making the Moon Full while he was in the *Selva Oscura*, and so he describes it as occurring on *Thursday* instead of *Monday* in Holy Week. Giambullari's words are 'Il Poeta nientedimeno, *per servirsene forse al senso mistico*, dice ch' ella fu tonda la notte che si ritrovò nella selva, laquale . . . fu la notte che è tra il Giovedì ed il Venerdì Santo.' This at any rate implies that Giambullari held to the view that the Vision commenced on Good Friday, April 8th.

[1] Since writing the above, I have been able to meet with a copy of this now very rare work. It is entitled *Del Sito, Forma, et Misure dello Inferno di Dante.* As this title indicates, it is mainly concerned with the topography of the Inferno and the measurements of its several parts ; following the lines of, and supplementing, the larger and better known work of Manetti with the same title. The passage in the text, though correctly cited by Dionisi, occupies a very subordinate position in the work. It occurs on p. 26.

Dionisi adds that the discovery by Giambullari of this apparent discrepancy between Dante's language and the facts of the case attracted much attention, but that his solution was not generally accepted. Mazzoni [1], Pelli, and others preferred to suppose that Dante's vision commenced on the evening of Monday in Holy Week [2], in order that the day spent in Paradise should fall on Easter Sunday. It appears to me that this apparently appropriate feature in the scheme depends on the mistake made by the Commentators of not noticing that Giambullari, while correctly stating that the actual Full Moon occurred (note the unusual expression) *fifteen hours after midday on Monday*, departed from ordinary usage in speaking of that hour as part of *Monday* at all, whereas it was clearly 3 a.m. on *Tuesday*, April 5th. Since then it is at any rate clear that Dante did not enter the Inferno till the evening *after* the Full Moon, whenever it was, this would be the evening of Tuesday and not of Monday, and consequently the passage through Paradise would not fall on Easter Sunday at all, but on Easter Monday.

Lombardi [3] appears to me to have fallen into a precisely similar error, in making the commencement of

---

[1] Mazzoni's treatment of this subject is quite 'a curiosity of literature.' See Supplementary Notes.

[2] *Anedd.* iv. p. 46.

[3] See his note on *Inf.* xxi. 112, when he introduces the above view with the words ' viensi per le vie additateci dagli astronomi a rilevaro che,' &c., as above.

Dante's journey (*i. e.* the wandering in the *Selva oscura*) fall in the night between Monday and Tuesday in Holy Week. He observes that Dante did not therefore compute the Anniversary of the death of Jesus as falling on the Good Friday of that year, but on the Tuesday, April 5th, this being the day *following* the Paschal Full Moon, and therefore (?) the day on which Christ was actually crucified (Come dal Vangelo si raccoglie . . . . nel giorno seguente al plenilunio antedetto). In other words, he seems to adopt a theory something like that suggested by Philalethes, that the actual day of the Crucifixion is fixed by Dante according to the Jewish method. But he puts the Full Moon (misled apparently by the language of Giambullari) on the *fourth* instead of the *fifth*, yet by the compensating error of placing the Crucifixion on the day *after* the Full Moon, instead of on that of the Full Moon itself (*Exod.* xii. 6, &c.), his ultimate conclusion agrees with that of Philalethes.

Dionisi himself speaks with some hesitation. At one time (p. 69, &c.) he appears to maintain strongly that March 25 was the day of Dante's entering the Inferno, pointing out that that day actually did fall upon a Friday in the year 1300. At the same time he throws out an alternative suggestion (p. 70) as tenable, though less probable, that Dante may have taken the true Equinox, which he puts at 4.40 p.m. on March 12th in the year 1300, and then have carried back to this corrected date the day both of the Annunciation and Crucifixion, and by consequence that of the commencement of his Vision. This is a series of the most extravagantly improbable hypotheses, as he seems to be himself a little conscious.

He finally defends the view which he prefers against what he calls 'la sola obbiezione che si possa fare' by the consideration that the Vision though 'maravigliosa e quasi divina, è pure fittizia' (p. 75), and consequently that Dante may not have troubled himself about the actual days of the month, or of the week, or of Full Moon, or of Easter in 1300; but assumed everything to be then in a typical condition: the year was the year of Jubilee; the season that of the Spring Equinox; the Moon Full[1]; the Ecclesiastical Year culminating with the Festival of Easter; and so on; all this being irrespective of any idea of piecing these general ideas together into a consistent whole of actual occurrence at any given time. The answer to this would seem to be :—If so, why has he gratuitously puzzled and confused us by a series of minute and precise, yet altogether meaningless and misleading, data of time, such as those we have already noticed, besides many more, and even more minute, references in the *Purgatorio*?

I pass over rapidly most of the theories to which

---

[1] See p. 73: 'Se non fu piena la Luna allora, lo fu nel dì della sua creazione; che Dio certo mostrolla tutta illuminata dal Sole.' See further on, p. 31, note 2.

these difficulties have given rise, such as that the Vision commenced (*i. e.* the wandering in the *Selva oscura*) on the night between April 2 and 3, *i. e.* on the eve of Palm Sunday (Gregoretti). So also Ponta, in his Appendix to the *Princ. Alleg. della Divina Commedia*, p. 227, where he starts from the strangely inaccurate statement that the Paschal Full Moon in 1300 was on Palm Sunday, the 3rd day of April! Or again between April 3 and 4, that is on the evening of Palm Sunday itself (Torricelli): or between the 4th and 5th of April (Arrivabene and Philalethes, though the latter doubtfully): or that it lasted from the 4th to the 16th of April (Lanci). Capocci again makes it begin on the night of Palm Sunday April 3rd in the *Selva oscura*, and extend to the end of Easter Sunday, April 10th. These and other theories or guesses are given by Pasquini in his *Principale Allegoria della Divina Commedia* (1875), pp. 229–30, and also by Lubin in his very elaborate and exhaustive *Studi*, p. 360. Many of these views, in the absence of reasons given (as I have not always been able to consult the original authorities), seem to me to have nothing whatever to recommend them. I only mention them to show how generally and how severely the difficulty that we are discussing has been felt.

There seem, however, to be four views which either on their own merits, or else on account of the distinguished names by which they have been advocated, deserve to be considered as within the range of practical criticism.

They are these:—

I. That the Vision began on March 14th, or the day of the true Equinox; which is the view most in favour with the early Commentators on the *Divina Commedia.*

II. That entrance of the Inferno is to be placed on March 25th, *i.e.* the 'ideal' Good Friday, as is maintained by the distinguished modern Commentators, Fraticelli and Scartazzini.

III. That it is to be placed on April 5th, the date of the Jewish Passover. This is suggested, as we have seen, somewhat tentatively by Philalethes, who, however, generally offers three alternative dates in his notes, regarding none of the views as free from difficulty.

IV. That the entrance of the Inferno was on Good Friday, April 8th, but that all the lunar references are to the Astronomical or Real Moon on April 5th; this is held, with candid admissions however of the unsolved difficulties which it involves, by Lubin.

There are two of the passages to which I have referred, which must be kept continually in mind, viz. *Inf.* xxi. 112, which plainly states that it was then Easter Eve: and *Inf.* xx. 127, which as plainly declares that 'iernotte,' that is on the night between Thursday and Friday, the Moon was Full. These are the two cardinal points which must never be lost sight of; the two main conditions which any theory is bound to satisfy.

I. Now of the four views just enunciated, the first was adopted among modern writers (I believe) by Giuliani, but chiefly (as I have said) by some of the early Commentators, who generally dashed at a conclusion without looking beyond the passage in hand, just as the copyists of MSS. in their textual emendations, often limit their critical vision to the compass of a single line. Moreover, they wrote before the discrepancy between Dante's allusions and the actual position of the moon had been noticed or suspected, since this was first observed, as we have seen, in 1544. This theory may therefore be disposed of without waiting for the consideration of the catena of passages of which it took no account.

The only reason for the assumption of the date of March 14th or 15th was that that was regarded as the date of the true Equinox at that time, and this seems to suit a very rigidly literal interpretation of a single passage, *viz.* i. 38–40, taken in conjunction with the recognized tradition as to the day of the beginning of Creation [1]. This implies that we are to bind down the reference in the passage in question to the *actual day* of the Sun entering Aries (and that moreover astronomically corrected), instead of referring it merely to the season, *viz.* early spring, when the sun was still in that sign; which suffi-

---

[1] See further the supplementary note on this subject.

ciently answers to the expressions used by Dante. It is not worth spending time on the detailed examination of this theory. It is enough to point out that the days in question (*viz.* March 14th and 15th) were Monday and Tuesday in 1300, Tuesday and Wednesday in 1301, and that the age and position of the moon would be quite unsuitable to all Dante's references in either case: the moon being then about at her Third Quarter in 1300, and just after New in 1301 [1]. At the same time this view seems to be favoured by several of the older Commentators who say that the Vision commenced *circa mezzo Marzo*. So *Pietro*, the *Ottimo,* and *Benvenuto.* But how worthless such opinions are, may be gathered from the glaringly false statements with which they are connected, *e.g.* 'fu allora la Pasqua fra Marzo' (*Ott.*); or 'circa la metà di Marzo, nel Venerdì Santo: la Pasqua caddi allora in Marzo' (*Benv.*) [2]. We have seen that

[1] Moreover there is a noteworthy passage in *Par.* xxvii. 87 where Dante, *then himself in the sign Gemini,* describes the position of the sun thus:—

il Sol procedea
Sotto mie piedi un segno e più partito,

i. e. the Sun was one sign and something more distant from Gemini. From this it seems clear that the assumed date of the Vision could not be the true Equinox, as the Sun would in that case still be near the beginning of Aries, and so very nearly two whole signs distant from even the *nearest* point of the constellation Gemini.

[2] *Pietro's* comment on *Inf.* xxi. 112 may be added, 'Nota quod auctor ostendit in hoc capitulo Christum crucifixum fuisse in medio Martii in aetate annorum xxxiv, et hoc opus incepisse in medio dicti mensis mccc.'

Easter fell on April in both years. Did it never occur to these old writers to 'verify their references'?

II. Next as to the theory of Scartazzini and Fraticelli, *viz.* that the Vision must be held to commence on March 25th[1], as the actual traditional day of the Crucifixion, which happened also, as will be noticed, to fall on a Friday in the year 1300. Of course the obvious objection arises that the Moon was *New* just when Dante's references state that it was *Full.* Scartazzini in answer to this says, that this *plenilunio* was merely 'una finzione poetica alla quale fa piede la tradizione della creazione del mondo.' But, if I rightly understand this suggestion, it is that Dante has had regard to the 'ideal' and 'traditional' date, not only in respect of the Crucifixion, but also in respect of the Creation, and moreover in the case of the Creation of the Moon as well as that of Man; the Moon being of course presumably created as a Full Moon[2], εὐλογώτερον γὰρ ἢ ἄλλην τινα. Well, but if so,

---

[1] It may also be remarked that according to Florentine (and some other mediaeval) usage the year began on March 25th.

[2] On this see Dionisi quoted *sup.* p. 26, note. B. Latini seems to have held the curious view that the Moon was created *New.* See *Tes.* ii. 48 : 'Et sappiate che 'l primo anno del secolo si fu el primo giorno de la Luna. La Luna hebbe el primo dì di Aprile 10 dì,' &c. In another part of the chapter however he seems conscious of some anomaly in supposing the moon to have been created in an invisible condition. The Tenth Century Anglo-Saxon Manual already quoted, *sup.* p. 14, says (p. 4), ' On the same (fourth) day He placed the moon Full in the evening in the East together with shining stars in the course of the autumnal Equinox, and fixed Easter-time by the beginning of the Moon.'

the ideal *plenilunium* would occur on the *fourth* day
of Creation, *i.e.* on Wednesday, March 23rd, and not
on Thursday, March 24th, as would be required by
the expression *iernotte*, spoken, as we ⁻have seen, on
Saturday the 26th. And (as I argued before) granting that Dante had full liberty to form or assume
ideal dates to any extent, yet we cannot suppose that,
having done so, he would bring everything into confusion by distinct and specific statements throughout
the poem inconsistent with such original conceptions ;
statements too quite gratuitous and arbitrary, which
might just as easily have been made consistent, or
indeed omitted altogether if they served no definite
purpose.

III. Next as to the hesitating suggestion of Philalethes, that Dante assumes a different sort of ideal
Good Friday, *viz.* April 5th (which was in fact a
Tuesday in 1300), as being the day for the cele-
bration of the Jewish Passover in the year 1300 in
accordance with the date of the actual Full Moon on
that day. He clears the way for this by arguing
that neither March 25th, nor the actual date of Good
Friday in 1300, *viz.* April 8th, can be adopted consistently with the two crucial passages, to which I
have directed special attention. As to March 25th, he
says that the language of *Inf.* xx. 127 finally disposes
of that, since *gar keine vernünftige Deutung zulässt* ('it
admits of absolutely no intelligible meaning'): and
similarly it forbids the acceptance of the date April

8th, because 'jedoch trifft auch hier die Angabe des Vollmunds nicht zu' ('the reference to the Full Moon does not agree with this either'). So again he denies that the Saturday on which these words are spoken can be supposed to be either March 26th or April 9th:—'Nach der Angabe des 9 April hätte sonach der Dichter sich um einige Tage geirrt; bei der Annahme des 26 Marz ist aber die Sache noch irriger' (*i.e.* on the adoption of April 9th the poet would have made an error of some days, while on the supposition of March 26th the matter is made worse). It is perhaps hardly worth while discussing the improbable suggestion of Philalethes further, since he does not profess to attach much value to it himself. He rather throws it out in despair, as his language just above indicates, of being able to rest in any other solution that has been proposed.

IV. Finally we come to the view advocated by Lubin, which in itself I make no doubt whatever is the correct one, *viz.* that the commencement of the Vision is to be taken in the natural and obvious sense of Dante's words as occurring on Good Friday, April 8th, 1300.

Lubin however is in constant difficulties, since he holds that Dante refers throughout to the *real* Full Moon which occurred on ~~March~~ 5th; and there is no course open to him but candidly to admit that the poet has fallen into an error. He says of this undoubtedly correct astronomical calculation, ' Ciò non significa altro se non che Dante nella Commedia siasi ingannato, ponendolo (il Plenilunio) due giorni dopo ' ; and adds '*Come ciò avvenisse non è facile dire*[1].'

Now it appears to me very singular that Lubin should not have seen the simple solution to be that Dante followed the *Calendar* and not the *Real* Moon. This is the more remarkable, because he does contend on the very same page (*Studi*, p. 362) that Dante ' *che seguì le credenze populari*' would naturally follow the Calendar of the Church in respect of the date of the Equinox, and also of the observance of Easter. (This is of course in refutation of the dates March 14th and 25th, to which we have already referred.) Now surely the same reasons precisely would lead us to infer, that he ' followed the Calendar of the Church,' and not independent astronomical calculations, also for the date of the Paschal Full Moon. In all the passages however where the question of the Moon's position is involved, Lubin gives that of the '*real*' Moon, and explains the hours intended by Dante as calculated on that hypothesis. Consequently, though his *days* are, as I believe, all right, his *hours* are, I venture to think, all wrong by nearly two hours, so far as they depend on indications of the *Moon's* position.

It is now high time to state distinctly and maintain

[1] *Studi*, p. 362, note.

emphatically the central principle for which I am contending, the application of which I venture to think dispels at once (if it is not too bold to say so) these clouds of doubt and difficulty as to Dante's language and meaning in his various allusions to time. I could scarcely enunciate it better than by a slight adaptation of the language of Lubin just quoted. 'It is natural to suppose that' (when referring to the Moon) 'Dante should have followed the Calendar of the Church.' In other words, all his allusions are, I believe, to be connectedly and consistently explained as referring to the *Calendar* and not the *Real* Moon.

Let me ask attention to these considerations.

(1) It should be remembered that Dante is not describing the scene as an eye-witness at the time, but is relating it some years afterwards, and moreover he is not describing an actual scene at all, but a purely fictitious and imaginary one, to which however he artistically imparts definiteness and reality by fixing very accurately the surrounding circumstances of time and place in which it is supposed to occur. Surely then his obvious course would be to make his references square with the computed position of the Paschal Moon as taken from the Calendar, since this would be the only available source of information for any ordinary reader, who might wish to follow him closely, and realise for himself, by interpreting the data so carefully and pointedly given to

him, the actual surroundings of each scene depicted.
Consequently, if Dante, when writing, wished to indicate the hour of Moonrise, say, on April 11th, he would
adapt his reference to the supposition that the Moon
was four days past the Full, as any one would find it
recorded in the Calendar for the year. Remember
too he is referring not to an ordinary Full Moon, but
to the *Paschal Moon* of the year, information as to
which would be universally and easily accessible;
the '*terme paschal*' being almost as conspicuous a
landmark in the Calendar of any year as the date
of Easter itself.

(2) Moreover let us ask ourselves, *why* is Dante so
careful to insert these various and frequent references
to time, not less I think than forty in the first two
Cantiche? We may be sure that they are not mere
poetic adornments, mere fixtures, so to speak, of
Dante's poetic furniture. All know doubtless how
one of the most characteristic and distinguishing
features of Dante's poetry is its extraordinary minuteness of detail in local description. Macaulay, in
his Essay on Milton, has compared Dante's descrip-
tions to the reminiscences of a traveller. Ruskin says
they often resemble the notes of a land-surveyor.
I believe that for vividness of effect he wished his
readers not only to follow him step by step in the
scenes which he depicted, but also hour by hour.
Consequently, to have adopted any other than a
popular computation of familiar celestial phenomena

would not only be poetically superfluous, but positively misleading[1].

(3) There could scarcely be a better proof of the inutility of adopting any other than the popular method of referring to astronomical phenomena, than the fact that it does not seem to have occurred to any one before Giambullari, *i. e.* for nearly two and a half centuries after Dante wrote, to go into the astronomical calculations. One is reminded of the familiar argument that the Bible uses the language not of science but of popular usage, because had it done otherwise it would have been unintelligible to those to whom it was addressed, and many generations must have passed before its true meaning could be ascertained. It seems that a similar fate would have befallen the work of Dante, had he corrected his astronomical references by independent scientific calculations.

(4) The strongest argument however would be *if* this hypothesis alone should give a consistent ex-

---

[1] A curious illustration of the difference between the ' Real ' and ' Calendar ' Moon, and of the necessity for any one writing for ordinary people, to keep clear of refinements of this kind, may be drawn from the phenomena of the year 1301, which some would assign as the year of the Vision. The *Calendar* Full Moon in that year was on Monday March 27th. Consequently Easter was kept on the following Sunday, *viz.* April 2nd. But the *Real* Full Moon was three days earlier, *viz.* on Friday March 24th (*auct.* Griou, Lubin, &c.). Hence we have the curious result, that if regard had been paid to the *Real* Moon, Easter Sunday ought to have been kept on March 26th, and not, as it actually was kept, on April 2nd.

planation of the various time-references in the poem, and this I must now endeavour to show. At any rate it is pretty generally admitted that other hypotheses do not do so, even by those who advocate them.

On the assumption then that this would be a natural supposition, I have constructed for myself a specimen of the sort of working Calendar by which Dante is likely to have guided himself, so that his time-references might be approximately correct, and also (what would be quite as important) popularly intelligible to his readers. As the Calendar Full Moon fell on April 7th, and further, as we learn from *Inf.* xx. 127, during the night between April 7th and April 8th, when Dante would (poetically speaking) have so observed it during the night spent by him in the *Selva oscura*, we should not be far wrong in supposing that it set on the following morning (*i.e.* Good Friday), about Sunrise, or within at any rate ten or fifteen minutes of the time of Sunrise. We should then be able to calculate, in such a rough and popular way as would be sufficient for Dante's poetical purposes, its rising and setting for the next few days, by allowing a retardation behind the sun of twenty-five minutes for each twelve hours or fifty minutes for each complete day[1]. Further,

---

[1] It is interesting to see Vellutello working out this kind of calculation for himself, allowing a daily retardation of about an hour, in his note on *Purg.* xviii. 76. Moreover, since writing this, I have found a passage in which Dante's own Master, Brunetto Latini, gives precisely this very rule for ascertaining practically the Moon's position on any

we should not be many minutes wrong if at a distance of about twenty days from the Calendar Equinox we assumed Sunrise to be about 5.15 a.m. and Sunset about 6.45 p.m. We could thus calculate roughly the impression that would be likely to be conveyed to an ordinary contemporary reader by a reference to Moonrise or Moon-setting during any of the days mentioned in the poem; and as the poet is surely likely to have used terms with a view not to the minute calculations of astronomers, but such as would be 'understanded of the people,' it seems to be most natural to suppose that he adopted some such rough and ready calculation as is here suggested. To go further than this would be (as Metastasio says of the too rigid application of rules like 'the Dramatic Unities') 'confondere il vero col verisimile.'

In fact Dante might well have reasoned with himself as to such scientific calculations in the language of 'il Filosofo,' whom he so often quotes :—

λέγεται περὶ αὐτῆς καὶ ἐν τοῖς ἐξωτερικοῖς λόγοις ἀρκούντως ἔνια, καὶ χρηστέον αὐτοῖς .... τὸ γὰρ ἐπὶ πλεῖον ἐξακριβοῦν ἐργωδέστερον ἴσως ἐστὶ τῶν προκειμένων.

Here then is the sort of working Table he might have followed :—

given day. See *Tes.* ii. 49, *init.*: 'E poi che l' uomo sa ciò (*viz.* the sun's position) e' può leggermente sapere ov' è la luna, chè ella si dilunga ciascun dì dal Sole tredici gradi, poco vi falla.' Since 13° of space are equivalent to 52 minutes of time, this is just the way of finding the moon's position that we have been describing.

April 8th, Friday evening,	Moonrise (retardation behind Sunset about 45 min.) say		7.30
,, 9th, Saturday evening,	,, (retardation about 50 min. more)		8.20
,, 10th, Sunday evening,	,,	,,	9.10
,, 11th, Monday evening,	,,	,,	10.0
,, 12th, Tuesday evening,	,,	,,	10.50

I am quite aware that these hours will not be found to correspond with those gathered from a modern Almanac for the Moon's actual risings and settings under similar circumstances in the month of April. The Lunar Motions may be well described as 'an excellent mystery.' The fact is that the difference of 50 minutes of daily retardation is merely a rough and average computation. It is subject to great variations at different times, and notably at the time of the Equinoxes. We have of course the well-known phenomenon of the Harvest Moon *rising* at nearly the same hour for several nights at the Autumn Equinox, and a corresponding, though less generally familiar, irregularity in the Moon's *setting* at the Spring Equinox. Still, even in this nineteenth century I imagine that most ordinarily instructed people are unaware (except in a very general way in respect of 'the Harvest Moon') of any difference of the Moon's motions at one time of the year or another, and that it is commonly supposed that the Moon rises and sets regularly about 50 minutes later every day. In any case I take it that Dante's critics are implicitly at any rate agreed upon this, that whatever *initial* assumptions be made as to Astronomical or Calendar Moon, a fixed retardation *per diem* is to be taken as the basis of subsequent calculations; though some dark suggestions have indeed been thrown out about allowing for this Moon being the Harvest Moon in the southern hemisphere, and consequently in Purgatory! If the popular method then be adopted afterwards, why not for the starting-point, *i. e.* for the day of the Full Moon itself? It would seem that *ce n'est que le premier pas qui coûte.*

Finally, let us always remember that we are interpreting a poem, not examining a scientific treatise; and while we insist, on the one hand, in the case of such an habitually accurate writer as Dante, on the necessity of assigning a definite meaning and purpose to these astronomical references, yet we decline, on the other hand, to analyse them as if they were announcements in the Nautical Almanac [1].

We will now show how a working scheme such as that I have suggested fits in with the principal time-references in the poem, and chiefly of course such as depend on lunar phenomena, though for the sake of a connected view I will briefly notice in order all the passages which I have been able to find containing time-references generally.

We shall I believe find it to be the case, that in

---

[1] See an illustration of Dante's use of popular astronomical notions, in reference to the planet Venus noticed later (p. 65).

all those places which are sufficiently precise to admit of a definite inference, explanation becomes hopeless in reference to the condition of the Moon on March 25th, etc. As between the dates of April 5th and 7th, *i. e.* those of the Real and the Calendar Full Moon respectively, I would observe that the difference being only that of two days, and in respect of the Moon's position about $1\frac{3}{4}$ hours, one time would generally speaking, do almost as well as the other. Still, we shall, I think, find some quite crucial and decisive passages in which this is certainly not the case.

The first reference, after the beautiful description of Sunrise on the Friday morning in i. 16, 37, etc. (which we need not discuss further), is in ii. 1, where we have the fading daylight and the darkening air (*aer bruno*) of the evening of the same day, at the moment of the commencement of their journey; when, as we read in the line immediately preceding, at the end of Canto I,

> Allor si mosse ed io li tenni retro.

VII. 98–9. The next reference is, I think, in vii. 98–9, which involves no difficulty. It is then just past midnight (Già ogni stella cade), whereas they had started at sunset (Che saliva quando mi mossi). This marks the passage from the fourth to the fifth circle.

XI. 113. The next reference occurs in xi. 113. The descent from the sixth to the seventh circle takes place when the Constellation of Pisces is 'quivering on the horizon.' The rising of this Constellation, covering of course several degrees of celestial space, commenced about 3 a.m. and ended about 5 a.m. We may suppose that the time indicated therefore is roughly about 4 to 5 a.m. The reference in the next line to Ursa Major lying right upon the north-west line ('tutto sovra Coro'), will be found, I believe, precisely accurate in conjunction with the other phenomenon. Antonelli, *Studi Speciali* (p. 86), says that when the Constellation Pisces is rising in a north latitude of 32°, Ursa Major will be '*tutto* in quel lato, l' estrema del timone distando circa 40° dal Polo.'

XX. 125. The next passage of this class is found in the two lines immediately preceding the important statement about the Full Moon, so often already referred to in *Inf.* xx. 127, as they are just about to leave the fourth Bolgia of the eighth circle. The time is here indicated by the 'setting of the Moon beneath Seville,' *i. e.* in the west. The extreme west limit of the world being regarded by Dante and his contemporaries as the Pillars of Hercules, this boundary is variously expressed by him as Spain, Gades, the Iberus, Morocco, etc., just as the extreme east limit is the Ganges. We shall find a series of passages in the *Purgatorio* in which such language is used. The time then here indicated by Moon-setting would be about 6 a.m. (or a little after) for the

Calendar Moon, and from 7.30 to 8 a.m. for the *Real* Moon. The former is not only intrinsically more probable, but moreover I venture to think that this will be found on further consideration to be a crucial passage in its favour.

For (1) it is natural that the poet should indicate to us the point where a new day commences. But (if I may anticipate what I am presently going to draw attention to) Dante seems purposely to avoid all reference to the Sun in the *Inferno*; so that it would seem that in this passage he prefers to speak of the *Moon-setting* rather than of the *Sun-rising*, in order to indicate (as it would approximately) the commencement of another day in his pilgrimage. Hence the earlier hour, which the Calendar Moon gives us, would be antecedently somewhat more probable.

(2) But there is a very much stronger reason than this, and one which I think definitely and conclusively settles the question in favour of the earlier hour, and consequently of the Calendar Moon.

The next reference to time after this is an absolutely precise one, *viz.* the often-quoted passage in xxi. 112, where Dante states that it is exactly five hours earlier than the hour of our Lord's death, which took place just 1266 years ago on the previous day. Seeing now that Dante in the *Conv.* iv. 23 distinctly argues both on *a priori* and on *a posteriori* grounds that our Lord's death took place at the *sixth* and not at the *ninth* hour, *i. e.* at noon and not at 3 p.m., it can scarcely be doubted that we are in this passage to take five hours before 12, and not before 3—in other words, 7 a.m., and not 10 a.m. It matters therefore little to note that Dante has erroneously cited S. Luke in this particular; the Evangelist's statement about the sixth hour referring not to Christ's death, but to His promise to the penitent thief (see Luke xxiii. 43, 44). For we may safely employ here in regard to the *hour* of our Lord's death the argument of Castelvetro in reference to the *years* of his life in this passage, *viz.* that we must adopt the view *maintained elsewhere by Dante himself.* His language is, 'in questo luogo seguita la sua opinione, non quella degli altri.' It should also be added that the early Commentators are absolutely unanimous on this point, *viz.* that 7 a.m. is the hour indicated. So I find it distinctly stated in the very early *Chiose Anonime* (edited by Selmi), *Jacopo della Lana*, the *Ottimo*, the *Anon. Fior.*, *Benv. da Imola*, *Buti*, *Landino*, *Vellutello*, *Bargigi*, and *Daniello da Lucca*[1]. There is not one who even raises a doubt on the point.

Applying this result then to the passage immediately before us in xx. 125:—If 7 a.m. be the hour so pointedly indicated (observe) *in the next Bolgia after this*, *viz.* the fifth, it would follow necessarily that we

---

[1] So also says Giambullari in the rare work already quoted (p. 98)— 'Certo è che già era levato il Sole per *un' ora intera* la mattina del Sabato Santo.'

must adopt the *earlier* hour (*viz.* about 6 a.m.) in the present passage referring to the *fourth* Bolgia, since the later hour (about 7.45 a.m.) would be clearly impossible. In other words, we must take the time of rising of the *Calendar* Moon, and not that of the *Real* Moon. I think then that this may be claimed, as I said, as a crucial passage in favour of the *Calendar* Moon.

XXI. 112. The important passage which comes next, *viz. Inf.* xxi. 112, has already been discussed, first, in reference to its bearing on the year of the Vision, and again, in reference to the hour of the day indicated by it in connexion with the last passage. There is however another interesting point about it which I should wish to notice connected with the reading in l. 113, the variation in which may possibly have some curious connection with the interpretation of the passage in its bearing on the *year* of the Vision. I noticed the following singular variant in one of the Bodleian MSS., and also in a beautiful little MS. in my own possession, *viz.* :—

Mille dugent' *uno* con sessanta sei.

I very much regret not having noticed this curious variant soon enough to include it among the selected passages which I have examined in a large number of MSS. I find it recorded by Scarabelli as occurring in five of the nineteen MSS. examined by him, including at least two highly important ones, *viz.* the

Codice Landiano at Piacenza and the most celebrated
of the MSS. of the Marchese Trivulzio at Milan,
these bearing the dates 1336 and 1337 respectively,
and being probably the two oldest dated MSS. in
existence (omitting one or two whose dates are either
probably spurious or obviously erroneous). This
reading also appears in the Commentary of Della
Lana (written in 1328), and is implied in the Commentary known as that of the 'False Boccaccio,' and
also I think in the important Commentary of Benvenuto de Imola. How is this curious variant to be
explained ? It is clearly I think spurious, since the
clumsy way in which the required unit is supplied
indicates a manifest after-thought, the rhyme no
doubt forbidding the obvious course of altering *sei*
into *sette*[1]. One can only guess, but there are three
possible suggestions that occur to me.

(1) It may have some bearing on the controversy
before us, *i.e.* it may have been introduced by some
one who thought that the assumed year of the Vision
was 1301. It is doubtful indeed whether any case
can be found of the explicit recognition of any doubt
on this point in early writers, still it may of course

---

[1] The device adopted by the presumed interpolator here reminds
one of the timid half measure by which a similar operator betrays
himself in Arist. *Poetics* xxiii. § 4. Here the original reading was
evidently ὄκτω, but some critic having bethought him of two additional
instances which he adds to the text, timidly inserted πλέον before ὄκτω,
instead of boldly altering ὄκτω into δέκα.

have occurred independently to an individual copyist at some early date.

(2) A second suggestion is that it may be due to the variation of a year which occurs frequently in the dates of old writers on account of the variation of the day when the year was thought to commence. It was sometimes Jan. 1st, sometimes Dec. 25th, and sometimes March 25th. Not only that, but those who agreed in taking March 25th differed as to whether the counting should start from the March 25th preceding our Lord's birth (*i.e.* the actual day of the Annunciation), or the March 25th *following*, *i.e.* the first occurring during our Lord's lifetime. The former was the view maintained by Dionysius Exiguus in the 6th century, and is sometimes said to have been followed in Pisa till about 150 years ago. The latter is said to have been the custom in Florence, at any rate about the 10th and 11th centuries. These statements are taken from the *Encl. Brit.* (*s.v.* 'Chronology'). They seem however to be at variance with the very ancient *Chiose Anon.* (*Ed.* Selmi), who says, 'Noi Italiani, *se non Pisani* faciamo menzione quando Christo incarnò nella Vergine Madre[1].' It should also be observed that Cacciaguida, when giving the date of his own birth, says that it was so many years

> Da quel dì che fu detto '*Ave*'
> Al parto in che mia madre, &c.,

[1] See Supplementary Note on Vedovati's Esercitazioni Cronologiche, p. 131.

following no doubt the Florentine use. (*Par.* xvi. 34.) Moreover the same Chronicler is not always consistent in his practice, probably in consequence of the different localities where his original authorities were found. Several of the Commentators on Dante notice this doubtful point, and maintain that in order to make thirty-four years in full for our Lord's life, we must date from the Conception, not the Nativity. Castelvetro, in his Commentary recently recovered and published, denies this, and prefers (characteristically) to say that Dante made an erroneous computation. It is at first a little confusing in old chroniclers to find the election of a Pope occurring apparently some months before the decease of his predecessor, as *e. g.* where Giov. Villani states that Honorius IV died in April 1287, and his successor Nicholas IV was elected in February 1287[1]. There is yet another source of confusion here, since a writer in giving a date sometimes refers to *anni compiuti*, and sometimes includes the *anno corrente* in his number. Possibly some of these various sources of confusion may have given rise to the variant under our notice.

(3) There is yet a third possibility. There were different theories as to the length of our Lord's life. Some maintained that he was in his 33rd year when

---

[1] It should be noted that this difference of a year would be liable actually to occur in the case of those who, like the majority of the early writers on Dante (as we have seen), referred the date of his Vision to the time of the true Equinox, '*circa mezzo Marzo*.' Compare the illustration from English History already cited, *sup.* p. 7.

E

he died, and others (as Dante himself) that he was in his 34th. Benvenuto da Imola, in his note on this very passage of the *Inferno*, says that he has known this point give rise to the fiercest contentions. It is possible therefore that this consideration may have had some bearing on the alteration in the text[1].

XXIX. 10. The next reference, in xxix. 10, presents no difficulty: *viz.* ' Now the Moon is beneath our feet.' This is another way of saying that it was early in the afternoon, about 1 or 2 p.m. Dante very significantly here, as in xx. 125 and elsewhere, avoids all mention of the Sun during his passage through the Inferno, and describes the hour by referring rather to the position of

> La faccia della donna che qui regge. (*Inf.* x. 80.)

XXXIV. 96. In marked contrast with, and only in apparent exception to this, is the next and last allusion of this kind which occurs in the *Inferno*, *viz.* in xxxiv. 96,

> E già il sole a mezza terza riede.

It will be observed that Dante and Virgil have now passed beyond the central point of the earth, and

---

[1] There is a curious passage in Sir John Maundeville (p. 77, ed. 1866) where he asserts that our Lord died at the age of 33 years and 3 months, but maintains that David was correct notwithstanding in prophesying that he should be forty years on earth in the words *Quadraginta annos proximus fui generationi huic* (quoting thus *Ps.* xcv. 10), because David referred to the old year of ten months, the other two months being added later by ' Gayus that was Emperor of Rome.'

have entered the Southern Hemisphere: and they have therefore quitted the Inferno, though they have still ' via lunga e cammino malvagio' to traverse before they come to the surface of the earth. As to the expression *mezza terza*[1], it occurs again in *Conv.* iv. 23 in connexion with *mezza nona* and *mezzo vespro*; and the time here indicated by it is obviously 7.30 a.m.[1] I should perhaps have noticed the words of Virgil (which speak for themselves however), in l. 68, 'Ma la notte risurge,' from which it appears that it was the commencement of night just before, whereas it is now suddenly about the same hour in the morning. This rapid change of twelve hours, to which Dante calls attention in lines 104–5,

> Come in sì poc' ora
> Da sera a mane ha fatto il sol tragitto?

is of course due to the passage through the centre of

---

[1] Some Commentators, as I find since, would not admit that this is ' obvious.' Pasquini makes the strange mistake (for such it must surely be) of interpreting *mezza terza* to mean 1.30 a.m., ' *un' ora e mezzo del mattino da mezza notte in su* ' (p. 259). So too apparently Benassuti, in his *Commento Cattolico*, ' Appendice ' to *Inf.* xxxiv, and in the note to *Purg.* i. 15, though not at *Inf.* xxxiv. 96. For having fixed the time then indicated as 2.50 a.m., he argues that as Dante had left the centre of the earth at 1.30 a.m., he had taken 1 hour 20 minutes in traversing the distance ! Benassuti also throughout adopts the astounding notion that when Dante passed to the Southern Hemisphere he found that *April* 10 had changed into *October* 9, which he says would be the corresponding day at the Antipodes ! This assumption appears repeatedly in the elaborate Tables by which his work is accompanied.

the earth, referred to just before in lines 88–93, as is
fully explained by Virgil in lines 106–118.

There is also one more rather doubtful allusion to
time sometimes quoted from *Inf.* xxxi. 10, as they
are approaching the ninth circle. It does not seem
at all certain whether it refers to the twilight hour
of the day, or to the permanent gloom of the place.
I certainly have always taken it in the latter sense.
This is advocated by Scartazzini, and among the
older Commentators, *Jacopo della Lana*, and the *Ott.*
take it thus ; but *Buti, Anon. Fior., Vellutello*, and
Daniello explain it of the hour of the day. Indeed
Vellutello takes occasion to collect in his note here
all the allusions to time throughout the *Inferno.*
Daniello very aptly quotes Virg. *Aen.* vi. 270 :

> Quale per incertam lunam sub luce maligna[1].

I certainly never myself understood this passage as
conveying a note of time when reading it independently of the present question. The chief objection
to doing so appears to be, that it would not allow
much more than *half-an-hour* for the very important
ninth Circle of the Inferno, with its four divisions
and its numerous episodes. I would also add that
l. 37, where Dante speaks of *forando l' aura grossa
e scura*, implies that the gloom was due to the quality
of the air and the nature of the place rather than the

---

[1] The preceding context in Virgil makes this citation still more
appropriate.

time of day. (Compare the language of *Inf.* iii. 29,
30.) I have sometimes fancied that Dante may have
had in his mind *Zech.* xiv. 6, 7, where we read of that
'day of the Lord' which is 'not day nor night'; and
it is certainly noticeable that in the Vulgate (which
of course Dante would have used) this language is in
connexion with the words (in verse 6) *Non erit lux
sed frigus et gelu* (translated quite differently in our
versions [1]), and that is precisely the case with the
scene on which Dante is here about to enter.

We now come to the curious and interesting question of the interval of time occupied in the passage
from the centre of the earth to its surface at the
Mountain of Purgatory. This was, as I have already
pointed out, about twenty-one hours, since in *Inf.*
xxxiv. 96 it is about 7.30 a.m., while in the last line
they emerge 'riveder le stelle,' and we learn from
Purg. i. 13–21 that these were the stars of early
morning (and therefore of course the following morning) with the 'Dolce color d' oriental zaffiro' already
in the sky, and Venus, the morning star, 'the har-
binger of dawn,' shining on the horizon, in other
words, about 5 a.m. or a little earlier. We may just
note in passing that it seems probable that a somewhat similar interval occurs in the passage from
Purgatory to *Paradise*, i.e. eighteen to twenty hours
seem to be unaccounted for [2]. As to the interval of

---

[1] This *var. lect.* is noticed in the margin of the Revised Version.

[2] I am aware that this is a disputed point. The view adopted in

twenty-one hours however in the passage between the *Inferno* and *Purgatory*, there can be no doubt[1].

the text, which is that of Philalethes and I believe most Commentators, involves the following explanation. The morning of the last day in Purgatory is described in that very exquisite passage at the beginning of Canto xxviii, after which follows the scene with Beatrice, the mystical procession, and the prophetic vision of the ills impending over the Church and shortly to burst upon her, including the captivity at Avignon. After that had passed away, the hour of noon is described (xxxiii. 103–5), and Dante is made to drink of the water of Eunoe. It seems to have been the following morning when he entered Paradise, as we learn from *Par.* i. 43—

Fatto avea di là mane e di quà sera.

In this passage *di quà* denotes the Northern Hemisphere where Dante is relating his Vision, and *di là* the Southern Hemisphere, that of Purgatory, which he was just on the point of quitting, and he goes on to repeat in different phrases that the latter was all light and the former dark. (*Purg.* xv. 6 is precisely similar, and *là* and *quì* refer respectively to Purgatory and to Italy. Compare again *Inf.* xxxiv. 118 where *quì* and *là* are again opposed, though the context implies the converse signification to each.) Della Valle thinks that the hour was not sunrise, but about 7.30 a.m.—(compare the '*mezza terza*' of *Inf.* xxxiv. 96)—having regard to *tutto bianco* in l. 44. This would be on the Friday in Easter Week (see his *App.* p. 61) according to his reckoning (to be explained later). It must be admitted that this unexplained interval is not free from difficulty. Mr. Butler contends that the passage from *Purgatory* to *Paradise* was instantaneous, and precisely at noon on the *Wednesday*; and that *mane* and *sera* in *Par.* i. 43 stand for the two periods, 6 a.m. to noon, and noon to 6 p.m., respectively. Benassuti also strongly maintains this view.

[1] Several Commentators have noted the propriety (?) of the descent and ascent taking approximately the same time: *e. g.* Giambullari (p. 151), 'Et avvenga che la ragione detti per se medesima che le distanzie uguali vogliono uguale il tempo del loro cammino: et conseguentemente tante ore voglia il risalire, quante ne ha volute lo scendere,' &c. See also Manetti on this subject quoted in Supplementary Notes. The descent, to speak accurately, took about four hours longer than the ascent, which Manetti gravely justifies in the passage quoted.

But now a curious point arises. What was the morning, the stars of which greeted Dante when he emerged on the surface from the bowels of the earth? Or in other words, when he passed that central point—

> Al qual si traggon d' ogni parte i pesi,

did he *gain* or *lose* twelve hours? Did this sudden transition (see again lines 104, 105) involve putting the clock *back*, or putting it *forward*? Was the 'mezza terza' of line 96, 7.30 a.m. on Easter *Eve* over again, or on Easter *Day*? And, by consequence, were the stars of the following dawn those of Easter *Sunday* or Easter *Monday*? Now modern astronomical theories will not help us here, and we must be guided by considerations of probability or propriety, as such considerations would be likely to present themselves to Dante. I cannot myself feel the smallest doubt as to the answer to be given to the above questions, though it is far from being the generally accepted one. The clock must I feel certain be put *back*, and not *forward* at *Inf.* xxxiv. 96, so that though the Easter Eve of the *Northern* Hemisphere was spent in traversing the Inferno, the 'mezza terza' immediately after passing the centre of the earth was 7.30 a.m. on Easter Eve of the *Southern* Hemisphere, and this second Easter Eve was most appropriately spent in the gloomy passage through 'the lower parts of the earth[1].'

---

[1] I find since that this is very clearly expressed by Giambullari (p. 152): 'Giunsero poco avanti l' Alba lo undecimo *nostro* giorno di

The contrary supposition involves what I cannot but call the monstrous consequence that Dante, in spite of his most keen appreciation of 'the fitness of things,' has represented himself as devoting the whole of Easter Sunday (of all days in the year!) to scrambling down the 'vellute coste' of Lucifer, and groping along the 'cammino ascoso,' the dismal and dark path from the earth's centre to its surface (see the language of *Inf.* xxxiv. 95-99, 133-4, etc.). So far as his great Vision is concerned, Easter Day would thus be a blank, and the poet himself on that day wholly shut from 'il chiaro mondo' (xxxiv. 134). Yet Della Valle, in a supplementary note (*App.* p. 35)[1], deliberately adopts this view, and distinctly maintains that Dante came to the earth's surface (*Inf.* xxxiv. 139), and entered Purgatory, on the morning of Easter Monday, and that the 'mezza terza' of *Inf.* xxxiv. 96 was 'della Domenica e non del Sabato.'

Dionisi again boldly maintains that the whole third day (without observing apparently that it was Easter Sunday) was spent in passing from the centre to the surface of the earth (*Anedd.* iv. p. 77)[2]. He

Aprile, *Giorno Pasquale in quello Emisperio*, benchè a noi fusse la notte che immediatemente va dietro a lo stesso Giorno della Pasqua.'

[1] 'Noi supponnemmo che il poeta la (*i.e.* the mezza terza del Sole) facesse *della domenica, e non del sabato.*' He carries this out consistently throughout. See the *Orario* of Dante's journey given on pp. 60, 61.

[2] His comment on *Inf.* xxxiv. 96 is—'Sicchè trovò la mane del *terzo* giorno, *ch' ei consumò intero* in salire alla Zona temperata australe à riveder le stelle.'

further maintains that the first day of creation dawned on the Southern Hemisphere, partly as being the seat of the terrestrial Paradise, and partly because of the phrase ' evening and morning ' used in Genesis (chapter i) by Moses writing in the Northern Hemisphere! But surely Moses would have adapted his description to the scene he was describing in the Southern Hemisphere, if the argument is worth considering at all. I think on the other hand that Dante would have been ready with some good *a priori* reasons why in passing from the Northern to the Southern Hemisphere we should go *back* and not *forwards* in time, though of course *we* are aware that it is simply a question of the East or West direction we might take in doing so. Can we not imagine his arguing that the Sun's first morning light on any day would shine on that Hemisphere

> Sotto' 'l cui colmo consunto
> Fu l' uom che nacque e visse senza pecca.
> *(Inf.* xxxiv. 115.)
> Là dove Gabriello aperse l' ali.
> *(Par.* ix. 138.)

where was the seat also

> Dell' alma Roma e di suo impero.
> *(Inf.* ii. 20.)

rather than on the ' mondo senza gente ' (*Inf.* xxvi. 117). Surely to reach *that* we must wander ' diretro al Sol,' and when *their* day dawns, it is the day that has, so to speak, already shone upon the more favoured hemisphere. To them in short

Redit a nobis Aurora diemque reducit.

But more than this, we are not altogether left to *a priori* conjecture, for in that strange and very difficult chapter in the *Convito* (iii. 5), in which the imaginary central cities of each hemisphere, *Maria* and *Lucia*, are described, whatever may be the meaning of these mysterious fictions, it is perfectly clear that the former represents some central and typical spot in the Northern Hemisphere, and the latter its direct antipodes in the Southern. Dante then describes 'come il Sole gira,' and makes him start on an ideal journey round the world beginning 'nel principio d' Ariete.' His aspect and elevation at *Maria* are described, and then he adds 'gira intorno giù alla terra, ovvero al mare, sè non tutto mostrando; e poi si cela, e comincialo a vedere *Lucia*.' It is surely a natural inference from this, that any given day (such as Easter or any other), would, so to speak, visit first any place on the Northern Hemisphere, and twelve hours later the antipodes to that place on the Southern Hemisphere.

There is another consideration which should not be overlooked. Though Dante carefully avoids any indication of the lapse of time in Paradise itself, yet we get one or two hints in some very obscure and difficult passages as to the lapse of time *meanwhile* on the 'aiuola' of this earth. From them it may be inferred that the journey through Paradise occupied only one day, *viz.* (as I should say) Thursday, April 14th, *i.e.* when Dante returned to the Earth again, it would have been found to be the evening of that day [1]. Thus the whole Vision would occupy precisely seven days. The completeness and propriety of such a period would not fail to strike one so constantly impressed by the mystic significance of numbers (see *Conv. Vita Nuova*, &c.) as Dante. If we suppose Purgatory to be entered on Easter Monday, we have the unmeaning period of eight days for the Vision.

It is only fair to add that I recognize the force of Dionisi's suggestion that the Earthly Paradise was in the Southern Hemisphere. Della Valle also lays stress upon this (*App.* p. 35). But this has always seemed to me a very curious imagination on Dante's part, and one full of difficulty if worked out to its consequences. It is needless to observe that few processes are more hazardous than that of crediting a person with even the logical (to say nothing of merely probable and conjectural) inferences that can in any aspect or connection be extracted from premises which he may admit. I have often doubted whether Dante would ever have maintained as a fact of serious belief this poetic fiction which was required for the purposes of his poem. How could he reconcile

---

[1] See further on this, p. 126.

the theory itself, for instance, with the general tradition of the Church, as well as the all but universal
mediaeval belief that the Earthly Paradise was in the
far East[1]? It will be remembered that having regard
no doubt to the language of the Book of Genesis,
Dante describes the joint source of the Tigris and
Euphrates as having been seen by him in the Earthly
Paradise at the top of the Mountain of Purgatory
(see *Purg.* xxxiii. 112–4). Or again would Dante
have ventured to maintain (except as a convenient
poetic fiction) that even *Purgatory itself* was in the
Southern Hemisphere, or on the Earth's surface at
all? This too was directly in contradiction with
orthodox tradition and teaching, including that of
Dante's master and guide, S. Thomas Aquinas, who
describes Purgatory as ' locus inferno conjunctus sed
superior eo.' Further, S. Thomas connects it locally
with the *Limbus Infantum* and the *Limbus Patrum*, from
which it is entirely separated in Dante's system.
The explanation suggested by Scartazzini in his
interesting notes on *Purg.* i. 1 and vii. 4 is probably
the correct one, *viz.* that Dante was compelled to

---

[1] See B. Latini, *Tes.* iii. c. 2 : ' In India e il Paradiso terreno.'
(India is of course here and elsewhere a general name for the East.)
The same idea occurs in almost any of the geographical or scholastic
writers cited elsewhere in these notes. Comp. also Fazio degli Uberti,
Dittamondo I. c. xi, where he makes his guide (Solinus) say on this
subject :—

> Diverse opinioni
> State vi son, ma suso in Oriente
> Per la più parte par che si ragioni.

invent something different from the gloomy conception ordinarily in vogue, in order that he might
represent Purgatory as 'più poetico, più chiaro, più
luminoso, più lieto, più ridente [1].'

I will conclude this part of the subject by drawing
attention to a characteristic which marks all these
indications of time in the *Inferno*, so far as they depend on references to the position or movements of
the heavenly bodies, and to the very marked and
noteworthy contrast between those that occur in the
Inferno and *Purgatorio*. The Sun of course would not
be visible in the *Inferno*. His absence is more than
once pathetically referred to, as in vii. 122, where the
upper world is by contrast described as

> Nel' aer dolce che dal sol s' allegra;

and in xxviii. 56, when Dante is addressed as

> Tu che forse vedrai lo sole in breve.

But whether visible or invisible, it is surely not without purpose that the Sun is never even once referred
to as affording a datum of time, the gate of the Inferno
once passed. Thus when Dante would indicate an
hour shortly after midday, we have seen that he does
so by saying that 'the Moon is already beneath our
feet' (*Inf.* xxix. 10). (It may be observed in passing
that with the assumed date of March 26th it would be
directly overhead.) When he would describe the approach of dawn, we have the cheerless indication that

---

[1] See a supplementary note on the traditions and beliefs referred to
in this paragraph, p. 122, &c.

'the fishes are quivering on the horizon.' Observe too the word 'quivering' (*guizzan*). He will not even here use a word implying light. Now contrast with this the passage where the same phenomenon is referred to in the *Purgatorio* (i. 19–21), and the same constellation is again rising on Easter morning. Then we are reminded of its light, though veiled, it is true, by the superior brilliancy of the planet Venus. We have light against light, as in a picture of Fra Angelico. Again in the *Inferno* even the time of Sunrise itself is indicated by the setting of the Moon, as we have seen in *Inf.* xx. 126. Here again it is most significant to contrast the language in which this very same incident of Moon-setting is presented to us in the *Inferno* and in the *Purgatorio*. In the former (see *Inf.* xx. 127) it is, 'Cain and the thorns are touching the wave beneath Seville,' thus making the allusion as unlovely as possible, and not even describing the phenomenon as present or visible, but merely as a fact marking time; nothing more than an astronomical datum. Now turn to *Purg.* x. 15, where the Moon-setting is again employed to furnish a note of time. Here it is no longer a bare astronomical fact that is described, but a sight vividly realised. Observe how the precise visible aspect of the Moon, which had only just disappeared, is noted; it is 'lo scemo della luna,' 'the Moon's diminished orb' (now in fact three and a-half or four days past the Full); and then we have the gentle and peaceful image

'Rigiunse al letto suo per ricorcarsi,' 'regained its bed
to sink again to rest' (Longfellow). *Before* the gate of
Inferno is reached, we have joyous and loving allusions to the bright Sun, 'Che mena dritto altrui per
ogni calle' (l. 18), and again the brilliant passage (in
lines 37, &c.) describing his rising at the beginning of
the Spring-tide as when he was at first created. And
then again in *Inf.* xxxiv. 96, the *very instant* Dante
and Virgil have passed out of the Inferno and have
entered the other hemisphere, and even before Dante
himself understands what has occurred, as though he
rejoiced to be again 'unmuzzled,' he at once makes
Virgil indicate the hour by a reference to the Sun

> E già il Sole a mezza terza riede.

Yet between these two limits he never forgets that
he is *dove il Sol tace* (l. 60); that he is in 'a land of
darkness as darkness itself, and where the light is as
darkness,' and he takes care that it shall be to us
'a darkness that may be felt.' But in the *Purgatorio*
all this is changed

> Uscendo fuor della profonda notte
> Che sempre nera fa la valle inferna.
>
> (*Purg.* i. 44-5.)

There he is careful everywhere to make us feel the
Sun's actual presence in his light and heat, and that
of the Moon 'walking in brightness.' We are constantly reminded how 'truly the light is sweet, and
a pleasant thing it is for the eyes to behold the Sun,'
and in his unrivalled descriptive power he secures
that we shall behold it with him.

Part II. *Purgatorio.*

In the Purgatorio we are able to follow the Poet's steps even more closely than in the Inferno by the help of the numerous indications of time which he has given us. The period occupied in traversing Purgatory is much longer, *viz.* four days, as against about 25 hours in the case of the Inferno. To speak more precisely, he is one day in the Ante-Purgatory, two days in Purgatory proper, and one day in the Earthly Paradise at the summit of the Mountain of Purgatory. There is no doubt about the aggregate amount of the time allotted, though there is, as we have seen, much dispute as to the day of the week or month on which the journey through Purgatory is supposed to commence, and of course very great controversy as to some of the hours indicated on the several days. The references to time are very numerous and minute. I have counted as many as thirty, but happily in regard to the large majority of them no question will arise. It will be interesting however to indicate them briefly in the order in which they occur, so that we may be able to understand better the general plan of the poem.

I. 19–21. Our first reference is, as has already been mentioned, i. 19–21, *i.e.* an hour or so before Sunrise on Easter morning, April 10th (as I have already maintained). The only point calling for notice here is the curious piece of hypercriticism on the part of some ingenious persons who have discovered by computation that ' Lo bel pianeta che ad amar conforta,' *i. e.* of course, Venus, was not actually a morning star in April 1300, but rose after the Sun. But it is evident that Dante wishes to describe the hour before sunrise under its most familiar, and so to speak its typical, aspect in the popular mind, and with that hour the brilliant Morning Star is generally associated. We may add too that if it *were* actually visible at that season, it would of course be associated (as Dante has with a realistic touch indicated) with the constellation Pisces, the Sun being in the next following sign of Aries. This is an illustration of the principle I have already contended for, that Dante in his astronomical allusions does not feel bound to sacrifice poetic effect, or a picture that strikes vividly on the popular imagination, to the exigencies of strict scientific (shall I not rather say, pedantic?) accuracy. You might as well object to a landscape painter that he had slightly altered the actual position of a tree or a house, as tested by results of mensuration or trigonometry. (It may be noted that precisely the same question arises in *Purg.* xxvii. 94–6, when the hour before dawn is again described in similar language.) Next in l. 107 we have the Sun on the point of rising, and in l. 115 occurs that exquisite picture of the breeze that precedes sunrise,—for that is most probably the

meaning of *ora* (compare *orezza* in xxiv. 150)—ruffling
the surface of the sea[1]. Next in Canto ii. l. 1, we
have the Sun actually upon the horizon.

II. 1-9. The next passage (ii. 1-9) is one of a
group of five or six similar in character, which may
therefore conveniently be considered together, since
the key to their explanation is the same. To understand them a brief exposition of the rude system of
geography adopted by Dante here and in the *Convito*
is necessary. It was in fact the same as was current
in his day, and its main features are to be found
in such writers as Orosius, Isidore, and Brunetto
Latini (with all of whom Dante was acquainted),
and in almost any of the old *Mappae Mundi*, such as
that of Hereford Cathedral, and many others which
are given by Lelewel in his *Géographie du Moyen Age*.
The habitable world was of course confined to the
Northern Hemisphere, the other was the 'mondo
senza gente' of *Inf.* xxvi. 117[2]. The Southern

---

[1] This wind of Daybreak, it will be remembered, forms the subject
of a beautiful little poem of Longfellow beginning, 'A wind came up
out of the Sea.' *Ora* in this sense is found in *Conv.* ii. 1—'Drizzato
l' artimone della ragione all' *ora* del mio desiderio, entro in pelago.'
It occurs also in Petrarch. Benassuti paraphrases—'quel venticello o
brezza, che sempre sentiamo alla mattina dal principio dell' alba.'

[2] In the following passage from the *Quaestio de Aqua et Terra*,
a dissertation publicly read by Dante 'inter vere philosophantes
minimus,' at Verona on Jan. 20, 1320, his views on this subject are
distinctly stated (see § xix): 'Nam ut comuniter ab omnibus habetur,
haec habitabilis extenditur per lineam *longitudinis* a Gadibus, quae
supra terminos Occidentales ab Hercule positos sitae sunt, usque ad
ostia fluminis Gangis, ut scribit Orosius . . . Per lineam vero *latitu-*

Hemisphere in fact contained no land except the Mountain of Purgatory, and the belief in the possibility of Antipodes would no doubt have been held, as by St. Augustine (*De Civ.* xvi. 9) to be unscriptural. The Northern Hemisphere was symmetrically divided into two parts, Asia in the east, and Europe and Africa in the west. Asia, in which Egypt was included (see *inter al.* B. Latini, *Tes.* iii. c. 2), was held to be equal in size to Europe and Africa together [1]; this being sometimes accounted for by the *a priori* consideration that it was the inheritance of Shem, the first-born, who had consequently a double portion! *e.g.* Gervase of Tilbury, *Otia Imp.* Dec. ii. § 2, 'Est Asia multo major quam Europa vel Africa : *Sem* enim qui eam obtinuit primogenitus erat, ideoque majorem et uberiorem partem accepit.' So also Fazio degli Uberti *Dittamondo* i. c. vi.

> Sem ebbe nome il primo, e' l suo dimoro
> In Asia fu, e quella parte tenne
> Ch' è grande per le due e ricco d' oro.

Europe and Africa were again symmetrically sub-

dinis, ut comuniter habemus ab eisdem, extenditur ab illis quorum zenith est circulus equinoctialis, usque ad illos quorum zenith est circulus descriptus a polo Zodiaci circa polum mundi, qui distat a polo mundi circiter xxiii gradus : et sic extensio latitudinis est quasi lxvii graduum, et non ultra, ut patet intuenti.' Thus the habitable globe comprised merely the temperate and torrid zones of the Northern Hemisphere, and only such portion of these as extended over 180° of longitude.

[1] This idea is found in B. Latini (*Tes.* iii. c. 1), Orosius, Isidore, Rabanus Maurus (*De Univ.* Lib. xii. c. 2), all of whom are mentioned by Dante, and others.

divided by the Mediterranean, which (as we learn
from *Par.* ix. 84–7) was regarded by Dante as reaching
half-way across the hemisphere, and thus extending
over 90° of longitude. His words are that from the
Ocean which surrounds the world

> Contra il sole (*i. e.* eastwards)
> Tanto sen va, che fa meridiano
> Là dove l' orizzonte pria far suole.

Jerusalem was in the system of Dante, as of the
other authors we have referred to, and indeed in
general mediaeval belief, the ὀμφαλὸς τῆς γῆς, and
this is therefore the 'Greenwich,' so to speak, of
Dante's computations of longitude, and consequently
of time [1].

On either side of Jerusalem, at the distance of 90°,
were the Ganges on the east, and the Pillars of
Hercules on the west, this limit being also variously
indicated by Dante as Spain, the Ebro, Seville,

---

[1] Apart from *a priori* reasons which Dante and others would probably have found in abundance to justify this symmetrical geography,
they would doubtless have considered the central position of Jerusalem
to be proved by *Ezech.* v. 5, *Haec dicit Dominus Deus: Ista est
Jerusalem, in medio gentium posui eam, et in circuitu ejus terras.*
Sir John Maundeville proves the same point by referring to *Ps.* lxxiii.
12, *Vulg.* (or lxxiv. 12 our *Bible* version), and he describes also a
simple method by which it can be proved experimentally! (p. 183,
Ed. 1866.) It is curious to note that Gervase of Tilbury, *Ot. Imp.*
Dec. i. § 10 (*fin.*), cites the same experimental proof in favour of
Jacob's Well, on which our Lord sat, being the precise central point
of the earth !

Gades, or Morocco [1]. Half-way between Jerusalem and Spain, and therefore in the centre of the Mediterranean, and at about 45° west longitude, was Italy [2]. Finally the direct antipodes of Jerusalem, and therefore at 180° either east or west longitude, was the Mountain of Purgatory. We find this distinctly expounded in *Purg.* iv. 67–71. (A reference to the Table and Diagram constructed to illustrate this at the end of the Book, will make this clear at a glance.) Now as 15° of longitude are equivalent to one hour of time, 45° = 3 hours; 90° = 6 hours; and of course 180° (as in the case of antipodes) = 12 hours difference of time.

It follows at once from this simple and symmetrical system of geography that if it be, for example, noon at Jerusalem it will be 6 a.m. in Spain (*i. e.* roughly speaking, Sunrise at the time of the Equinox); 9 a.m. in Italy; 6 p.m. (or Sunset) in India; and midnight

---

[1] Every one will recollect Juvenal's expression of this popular geography in the lines—

    'Omnibus a terris quae sunt a Gadibus usque
    Auroram et Gangem.'
              (*Sat.* x. 1, 2.)

See also *Inf.* xxvi. 107-11, and Petr. *Son.* clvi.: 'Non dall' ispano Ibero all' indo Idaspe.'

[2] To judge from some of the Mediaeval *Mappae Mundi*, I imagine that Babylon was regarded as being 45° east of Jerusalem, as Rome was 45° west of it. I have not been able however to find any direct statement of this in words, and certainly there is no recognition of this temptingly appropriate bit of symbolism in Dante himself, as one might have expected, had it been generally recognized.

in Purgatory [1]. With this key the five passages which I have referred to will, I think, become clear at once. They are as follows:—

(I.) ii. 1-9. The first three lines of this Canto describe, somewhat *per ambages* it is true, sunset at Jerusalem: it was consequently Sunrise in Purgatory, —'*Là dova io era*' (l. 8)—and midnight on the Ganges (l. 5): for night, here and elsewhere, when spoken of generally as being in any spot, naturally stands for midnight, as its central point. In passing, a word of explanation of the obscure lines 5 and 6 may not be amiss. The Sun being in Aries, the night, revolving exactly opposite to him (l. 4). is considered to be in Libra (*le bilance*), and the Scales are said to fall from the hand of night when night overcomes the day (*soverchia*), *i.e.* becomes longer than the day. This of course it does after the autumnal Equinox, and since the Sun then enters Libra, that constellation ceases to be within the range of night, and so the Scales are poetically said to fall from the hand of night [2].

---

[1] It is right to observe that these geographical, as well as the astronomical, allusions in the poem are worked with extreme minuteness of detail by Della Valle. My contention however here and elsewhere is that such details, however scientifically accurate, are somewhat superfluous, because, even if Dante as a student was acquainted with them (which we may sometimes doubt), Dante as a poet would not have regarded them.

[2] Of course when Libra ceases to belong to the hemisphere of night, Aries commences to do so. and this explains the curious expression in *Par.* xxviii. 117, 'Che *notturno Ariete* non dispoglia,' to describe the

(II.) Next we have iii. 25. We gather from ii. 55, and iii. 16 (which I shall mention presently), that it was now about one hour after Sunrise in Purgatory, say, if you will, about 6.30 a.m. It would consequently be 6.30 p.m. in Jerusalem, and, according to the above computation, about 3.30 p.m. in Italy, where, as Virgil here says, his body is buried. Dante's use of 'Vespero' is sufficiently explained by *Conv.* iii. 6 and iv. 23 to be the last of the four divisions of the day, *i. e.* from 3 to 6 p.m., which therefore corresponds exactly with its sense in this passage.

(III.) iv. 138-9. The third passage in this group is iv. 138-9, where it is midday in Purgatory (l. 138); and following out the same calculation as before, it will be midnight in Jerusalem, and consequently sunrise on the Ganges, and sunset in Spain or Morocco; and the hemisphere of night will consequently extend from the Ganges to Morocco. Now this is exactly what Dante means by saying that starting from the bank or river's edge (taking the reading *dalla riva*) night's advancing foot just falls upon Morocco, *i. e.* night is just commencing there. (The reading *ed alla riva* is, I imagine, simply a blunder resulting from a wrong division of the words.)

absence of Autumn or Winter in Paradise, where it is perpetual Spring (l. 116). Again the word *soverchia*, to describe the preponderance of night, is well illustrated by an expression in the Anglo-Saxon Manual which I have elsewhere cited. On p. 11 we read, 'It is needful for us to hold the holy Eastertide by the true rule, never before the Equinox, and *the darkness being overcome.*'

(IV.) **xv. 1-6.** We next have the rather obscure passage xv. 1–6. The first five lines express, with a good deal of circumlocution, that three hours of daylight remained in Purgatory—in other words, that it was about 3 p.m. Consequently it was *Vespero là* (in the sense of the word 'Vespero' already explained in the note on iii. 25), and *qui*, *i.e.* in Italy, where Dante is then writing or narrating, it was midnight [1]. This follows quite simply as before, thus:—Purgatory, 3 p.m.; Jerusalem, 3 a.m.; and consequently three hours earlier, *i. e.* midnight, in Italy.

(V.) **xxvii. 1-6.** Finally there is the passage in xxvii. 1–6, which is interesting partly from the completeness with which Dante goes through these calculations of synchronism, but still more from the variations of reading *nona*, *nuova*, and *nuovo*, in l. 4. These are instructive, because it is clear that the comparatively unusual word *nona* was not understood by the copyists, or at any rate they were all adrift in regard to its meaning as here employed. Consequently some read *nova*. This, being quite unintelligible, led to a further alteration *novo*, and then once more *da* was altered into *di*. This gave a grammatical sense at any rate, but when we come

---

[1] For other instances of a similar opposition of *qui* and *là*, see the passages already quoted on p. 61 *note*, and compare also *Par.* i. 55, ' Molto è licito là, che qui non lece,' &c., where the context shows that *qui* means here on earth, and *là* in the Earthly Paradise.

to attach a meaning to the words the result is a statement false and nonsensical, since it practically would describe sunrise, *i. e.* the light burning forth *di nuovo*, as taking place at the Ganges at the same time as at Jerusalem (see lines 1 and 2), which is manifestly absurd. The corrupt readings here (as is often the case) have a large majority of MSS. on their side, in somewhat the following proportions, according to the collations I have been able to make. I have found *nona* in 65, *nova* in 77, *novo* in 64 MSS. With the true reading *nona*, the interpretation proceeds quite simply as before. It was sunrise in Jerusalem (lines 1, 2); consequently midnight in Spain (l. 3); (note how Libra is used here exactly as in ii. 5 just explained, to indicate the middle point of night while the Sun is in Aries at the vernal Equinox). It was therefore noon at the Ganges (l. 4); and consequently (*onde*, as Dante concludes in l. 5) it was sunset, or the day was departing, in Purgatory.

I have taken these passages somewhat out of their natural order, but I thought it would be conducive to clearness to consider them together, and it will absolve us from the necessity of further discussion when we note them in their order. It is perhaps hardly necessary to add that passages of this kind occur only in the Purgatorio, since, unlike the Inferno and Paradiso, it is supposed to be a definite spot on the earth's surface, having its own latitude and longitude. Hence in the case of indications of time in this part of the poem, it must be clearly understood, what is the meridian referred to,

> Che qua e là, come gli aspetti, fassi,

as Dante reminds us in *Purg.* xxxiii. 105. I am glad also to have drawn special attention to these passages, in order to shew what a really simple and intelligible meaning they convey, because I know many readers of Dante act on the principle of

> Non ragionam di lor, ma guarda e passa

as soon as they find him embarking on astronomical or geographical subjects. I hope they will be convinced that in these cases Dante's own *dictum* in *Par.* xxviii. 60 applies :—

> Tanto, per non tentare, è fatto sodo.

II. 55. To return now to the consideration of the time-indications in order, we have next to consider ii. 55, where 'now 'tis perfect day,' and the Sun probably a few degrees above the horizon. If, as l. 57 states, Capricorn had cleared the meridian, Aries would have cleared the horizon. I don't know whether we need go into such a refinement as that of Della Valle, who observes, that the Sun being now 21° advanced into the sign Aries, and that constellation occupying 30° of celestial space, the Sun would be precisely 9° above the horizon, which would represent about forty minutes after sunrise, and if so about

6 a.m. Certainly it could scarcely have been more, since in the next ref., *viz.* iii. 16, the Sun is still flaming *red*, which phenomenon, as Della Valle observes (see Scartazzini's note), does not generally last (unless in exceptional conditions of that atmosphere, out of the question here) more than an hour after sunrise.

IV. 15. We turn next to iv. 15, when the Sun is up fully 50° above the horizon, *i.e.* about 3½ hours, or some 2½ hours since the last time given : in other words, the hour is from 8.45 to 9 a.m. Some Commentators have made a difficulty that this gives more than two hours for the colloquy with Manfred in the previous Canto, and they interpret the last ref. in iii. 16, as indicating two hours of daylight. But this supposition is both improbable from the reasons already assigned, and also unnecessary, since (1) Dante apologizes, so to speak, in the lines preceding this for the lapse of so much time unnoticed : (2) the whole of the intermediate time is not devoted to the interview with Manfred, since some time may probably have been lost in hunting for the road in which both Virgil and Dante are represented as occupied in iii. 52–57 : and also (3) after that, they are expressly stated to have walked a mile. At least this is certainly I think the most probable interpretation of iii. 68. It is distinctly so explained by Buti (who adds a mystical interpretation of the various details) and by Daniello da Lucca. The other older Commentators appear to pass over the passage.

However, the difficulty raised is at best a very trifling one.

IV. 138. This passage has been already discussed. It is then midday, thus implying another interval of about three hours. This has been occupied in the tedious and difficult passage of the narrow gorge, described in lines 22–34: the ascent of the 'balzo' above it, lines 46–51 : the rest and discourse which then followed, lines 52–99 : and the interview with Belacqua seated near a 'petrone,' l. 101, to which Dante and Virgil had dragged themselves, l. 103.

VII. 43 and **85.** The next allusions of this kind occur in vii. 43 and 85, in the former of which day is declining, and in the latter there is *poco Sole* left. This causes them to hurry on to the beautiful valley of the kings, a scene that must live in the memory of any one who has read the exquisite lines in which it is described.

VIII. 1. Then follows the celebrated description of the deepening twilight in viii. 1, &c., which every one knows as one of the most beautiful and touching passages in the whole poem, together with the singing of the compline hymn by the spirits of that 'noble army' (l. 22). The 'squilla' in l. 5 probably alludes (as *Scart.* suggests) to the *Ave Maria* shortly after Sunset. In viii. 49 it was getting nearly dark, but objects were still discernible if looked at closely, and so it was perhaps about 7. 30 p.m. We have now arrived, it will be observed, step by step, at the end of the first day in Purgatory, which, to speak more accurately, has been spent in the *Ante*-Purgatory. It is therefore now, as we have maintained, the Evening of Easter Sunday, April 10th.

IX. 1–9. We now come to that *crux interpretum*, the passage at the beginning of Canto ix, one of the most difficult and disputed passages in the whole of the Commedia, and one which I am afraid we cannot avoid discussing at considerable length.

If any one will refer to Scartazzini's exhaustive note at the end of this Canto, he will see that the literature of this passage might almost be described as a small library in itself. Further, if he will turn to the paragraph headed '*Risultato*' (p. 161) he will find that Scartazzini and others regard the difficulties as all but insoluble, since every interpretation suggested remains open to formidable objections. I do not flatter myself that I can altogether succeed where so many have failed. But I must venture to say that in working out the simple principle which I have been advocating, I was surprised to find how it seemed to me at any rate to clear up the chief difficulties of this celebrated passage, and to give a satisfactory and consistent explanation of its various details. The principle to which I allude is that Dante always refers to the Moon's age and position as it would be popularly understood, and as any one would find it recorded in the Calendar of the year, and that he does not take account of scientific cor-

rections of such popular views, whether he may have
had access to them or not.

Now following Scartazzini, or any others who have
discussed the passage at length, we may thus sum up
the chief points to be dealt with :—

I. Is the *Concubina del Titone antico* in l. 1 the *Solar*
or the *Lunar* Aurora? (It ought to be added that a
quite modern interpretation of Antonelli, adopted by
Scartazzini, changes the reading in l. 1, and denies any
Aurora whatever to be referred to. This will be briefly
noticed presently.)

II. What is the '*freddo animale che con la coda percuote la gente*'? (lines 5 and 6).

III. What are the '*passi con che la notte sale*'? (l. 7).

I. We will take these points in the reverse order.
Now I venture to think if, apart from the context, any
one were asked what is the most likely and obvious
meaning of 'the steps with which night ascends,' nine
persons out of ten would at once say the six hours
of the first half of the night, *i.e.* 6 p. m. to midnight.
Further, any person acquainted with the *Convito* would
be still more convinced of this by noticing what Dante
says in iv. 23, where he describes human life (following as he says 'il Maestro della nostra vita Aris-
totile ') as an Arch : so that our life is nothing else
than ' uno *salire* e uno *scendere*' (note the use of the
same word *salire*). He then proceeds to say that
the same metaphor may be applied to the year, and
also to *the hours of the day*. As to our life, the ' punto

sommo di quest' Arco' is put at thirty-five years, the 'mezzo del cammin di nostra vita'; and similarly noon is 'il colmo del dì,' and therefore no doubt by parity of reasoning, midnight would be 'il colmo della notte.' He then argues that for this reason our Lord willed to die in His thirty-fourth year, and also at the sixth hour of the day (quoting, not very accurately, St. Luke in proof of this), in order that He might not be associated with the decline either of life or of the day, 'chè non era convenevole la Divinità stare così in discensione.' (Note that last word again as in contrast with *salire* here. The bearing of this also on the interpretation of *Inf.* xxi. 112, already pointed out, will not be forgotten.) I think then we can scarcely doubt that the 'passi con che la Notte sale' are the hours (as I said) from 6 p.m. to 12, and consequently that the precise time indicated by the words which follow, *viz.* that two of these steps were already made and the third was now beginning to droop its wings (the metaphors are, it is true, a little mixed), would be shortly after 8.30 or between 8.30 and 9 p.m[1].

I should here add that other explanations suggested for *passi* are (i) *the watches of the night*—to which I object (1) the graphic word '*sale*' loses all force if applied indifferently to all the four watches: (2) the

---

[1] The two following passages should be noticed in which the *hours* (which we take *passi* to be here) are by a different metaphor described as the handmaids of the day: *viz. Purg.* xii. 80 and xxii. 118–120; the latter passage being in other respects something like this.

hour indicated would be about 2 a.m., which suits neither Solar nor Lunar Aurora. (ii) Another view is that the *passi* are *signs of the Zodiac.*—But (1) what definite idea can we attach in that case to the expression (and it is clearly meant to be very definite) 'the signs with which night ascends'? or (2) to the expression '*fare passi*' in such a relation? (3) As a matter of fact, those who adopt this view differ (as I might say both metaphorically and literally), *toto caelo* as to the signs which are referred to.

As to the expression '*nel luogo ov' eravamo,*' it is to be explained, I take it, in reference to such passages as *Purg.* ii. 1. 8 (where nearly the same words occur, 'Là dove io era'), and also the other similar places which we have lately been discussing, in which Dante notes that the time, when indicated by hours, or by reference either to the Sun's or Moon's position, is a variable term depending on the longitude of the place referred to.

II. What is intended by the '*freddo animale che con la coda percuote la gente*'?

Granting (as seems most natural) that some sign of the Zodiac is here referred to, I might again ask, without any reference to the general interpretation of the passage, which would any ordinary person think to be most likely? Clearly I think the *Scorpion.* Partly because that is the only one whose tail is conspicuous as an object of terror: partly because this description seems to be (as Dionisi, Scartazzini, and others have pointed out) a direct imitation or reminiscence of Ovid, *Fast.* iv. 163 :

> Elatae metuendus acumine caudae Scorpio ;

and *Met.* xv. 371 :

> Scorpius exibit caudaque minabitur unca.

I need not remind readers of the *Convito* how frequently Dante quotes the Metamorphoses of Ovid, sometimes by that title, sometimes as ' *Ovidio Maggiore.*' There are also many resemblances to passages in Ovid traceable throughout the *Commedia*, and notably in the earlier Cantos of the *Paradiso*, so that Mr. Butler even suggests that Dante had just been reading Books VII and VIII of the Metamorphoses when he wrote those Cantos. Indeed, as far as I know, no Commentator has ever thought of suggesting any other Zodiacal constellation except that of *Pisces*, which, though it would correspond with the Solar Aurora, and though it seems to suit the epithet *freddo*, is at once excluded by three considerations : (1) the singular *freddo animale* is out of place : (2) the reference to the tail in l. 6 becomes ridiculous : (3) there are no conspicuous stars in that sign, so that the beautiful description of l. 4, *Di gemme la sua fronte era lucente*, becomes unmeaning.

It should also be added that the constellation *Cetus*, or the Whale, has been suggested by some Commentators. This however is not a sign of the Zodiac, and it is much more probable (to judge from

many other similar passages) that Dante would have
referred to one of these for his Solar or Lunar data.
Also in any case *Cetus* can relate only to the *Solar*,
and not the *Lunar* Aurora.

But then I shall be asked, How about the epithet
freddo? How is it that the Scorpion is described as
'*freddo animale*,' especially in view (as it has been
urged) of Virgil's expressed *ardens Scorpio*.

I would reply to this :—(1) Virgil's epithet refers
specially to the constellation and not to the animal,
whereas in Dante the reverse is probably the case.
(2) The epithet *ardens* (as I believe most Commentators agree) describes not heat, but the 'burning
and shining light' of the brilliant stars in that Sign,
and notably of Antares. If so, its meaning corresponds with what Dante says in l. 4 just referred
to. As to the epithet *freddo* here, Philalethes suggests (what is probable enough if Dante were
thinking chiefly of the constellation and not of the
animal), that it is so called since it is associated with
November in the Sun's annual course. It is most
likely however that the epithet refers rather to the
animal itself, and if so surely the expression *freddo
animale* as descriptive of the Scorpion needs no justification. There are several lines of association
between it and coldness. First it is itself an inver-
tebrate, and consequently a cold-blooded animal :
further, its *habitat* is in cold and shady places,
under stones and in old trunks of trees, etc. : and

once more, its venom produces cold. As B. Latini says (*Tes.* v. c. 1), when speaking of poisonous serpents, ' Tutti i veneni sono freddi ' (he adds that *veneno* is so called 'poi chè li entra dentro dalle *vene!*')[1] Finally, I have met with the following convincing proof of the *naturalness*, so to speak, of the epithet *freddo* as applied to the scorpion in two passages in the Coltivazione of Alamanni, where he gives this precise epithet in one place to the constellation (possibly in the sense above suggested by Philalethes), and in the other case to the animal itself. Here are the passages:—vi. 281

> Quando al *freddo* Scorpion Delio ritorna.

And again in v. 1089, when he is giving an enumeration of the various pests to tender herbs, he mentions among others,—

> Il *frigido* scorpion, l' audace serpe, &c.[2]

I think then it may be laid down as quite certain that the Scorpion is the Constellation here spoken of as on the horizon, since it, and it alone, combines the three points here mentioned : (1) a brilliant group of

---

[1] Dr. Murray has kindly sent me the following apposite illustration from Ambrose Paré, one of the earliest of eminent French surgeons (1509-1590): 'Antonius Benevenius dit avoir eu un serviteur lequel fut piqué d'un scorpion, et tout subit lui survint une sueur *froide comme glace.*' Ricaldone's comment on *freddo* is curious. See *Suppl. Notes.*

[2] I have dealt with this point at some length, since it has been represented to me as a difficulty in my interpretation of the passage.

stars: (2) the suitability of the epithet *freddo*: (3) a formidable tail.

Let me note in passing a rather hypercritical objection quoted by Scartazzini, that the Scorpion does not *strike* with its tail but *stings*, and that Dante should have said *ferisce*, not *percuote*. This I think scarcely needs a serious reply, but if it does, I find πατάσσειν actually so applied in Greek, *viz.* in a passage of Apollonius, citing Aristotle to this effect.

III. Now lastly as to the first question. Does Dante refer to the *Solar* or *Lunar* Aurora? Both views have been vigorously contended for. Without going into details, I may observe at once that on merely *a priori* grounds many Commentators have argued for the latter, from the singular, and I may say unique, expression 'La *concubina* del Titone antico,' in l. 1, and I may add the choice of the word *amico* in l. 3 [1]. Aurora (*i. e.* the Solar Aurora) being according to the universal language of Mythology described as the wife of Tithonus, it has been inferred that Dante must have had a special reason for adopting the unusual expression *concubina* here, and that the word might not unnaturally express the sort of secondary position or inferiority of the Lunar as compared with the Solar Aurora.

And surely this would be a perfectly natural ex-

---

[1] As *Scart.* very well puts the argument (p. 154), 'Invece di chiamare la bella Aurora *moglie* o *consorte,* ei la chiama *concubina* di Titone; invece di dire costui *marito,* ei lo dice *amico* dell' Aurora.'

pansion of the mythological idea on the same lines. One feels almost sure that if the mythopoeic fancy had been called upon to account for the family relations of the Lunar Aurora, it would most likely have been done in this very way. Indeed this appears to me to be an ingenious and original touch very much in Dante's manner. To the objections of Scartazzini and others, that Dante would be *falsifying* mythology, I attach no weight whatever. It would not be *falsifying*, but rather *modifying* or *adapting* it. And indeed why should not Dante thus bend a *myth* to his purpose, just as he declares he himself was wont to deal with *words*? for he says that 'many times and often' (*molte e spesse volte*) he had made words say in his poems something different from that which they had been wont to express for other writers (Auct. *Ott. Commento*). I may also observe that the enormous majority of Commentators from the earliest times have understood the passage of the Lunar Aurora. (Among others may be named *della Lana, Benv., Buti. Land., Vell., Dan.*) Antonelli, who denies all reference to any Aurora at all, states an ingenious objection to the *Solar* Aurora at any rate, *viz.* that l. 4 would be inappropriate to it since its effect would be to quench or render insignificant the light of the stars. He quotes very aptly Virgil, *Aen.* iii. 521, as descriptive of this effect:

Iamque rubescebat stellis Aurora fugatis.

But however this may be, I think it is rather on

a posteriori grounds that the question must be settled :
I mean, which view best suits the passage ? I say
most unhesitatingly that this is the case with the
Lunar Aurora. For now let us examine the facts.
When I refer to Philalethes, I find that the Moon,
three nights after the Full, would by calculation rise
about 9 p.m., or a little after. Further I find from
the same writer, and also on the authority of Lubin,
who illustrates the astronomical position by a diagram, that the Moon was (as indeed will be evident
on a little reflection) within the sign of the Scorpion,
and though I really do not think we need trouble
ourselves with very precise calculations of her Right
Ascension, still it would, I believe, be about 16 hrs.
30 min., which would be just right. It will also be
remembered that the bright stars of that constellation
are those that rise first. So that it seems to me that
we have all the various details of Dante's description
fulfilled in the minutest particulars. For if I might
paraphrase the first few lines of the Canto in a
matter-of-fact and prosaic way I should read them
thus :—

The Aurora before moonrise was lighting up the
Eastern sky (lines 1–3) ; the brilliant stars of the sign
Scorpio were on the horizon (lines 4–6) ; and finally it
was shortly after 8.30 p.m. (lines 7–9).

All of which details we find to have been literally
the case if we imagine ourselves to be on the third
evening after Full Moon. But here we have a crucial

test as between the Calendar and Astronomical Moons, for as it was the evening of Easter-Day, April 10th, 1300, it was the *third* evening after the Calendar Moon of that year, but the *fifth* evening after the Real Moon. On the latter supposition, all the harmony of various allusions becomes discordant at once. Lubin, for example, haunted by his real moon, gives the period indicated as 10.30 p.m., and he naturally gets into difficulties with lines 7–9, and has to interpret *passi* as signs of the Zodiac. More-over the real Moon's Right Ascension would be fully eighteen hours, which would take her beyond Scorpio and within the sign of Sagittarius. Philalethes, firmly holding to the hour of 8.30, and also assuming the position of the real or astronomical Moon to be always taken by Dante, here as elsewhere, has to adopt a different day of the month, and then he can only make the allusions to Easter, &c., throughout the poem work, by the desperate resource of sup-posing Dante to be following not the Christian, but the Jewish calculations for the Passover. Mr. Butler objects to the statement that moonrise was about 9 p.m. or soon after, that the moon of the Spring Equinox would be the *harvest-moon* of the other hemi-sphere, and consequently of Purgatory, and so it would rise about 7 p.m. I must say I think this would be to demand from the poet an accuracy which would almost amount to pedantry. Had Dante said that on the occasion of some public event occurring at

a known day and hour, he observed, as he looked in
a particular direction, the moon passing behind the
tower of the Bargello, no doubt we should then be
justified in testing the words by a strict computation
of the actual position of the real moon at that hour.
But, as I have already said, since he was composing a
purely fictitious vision, at a time many years after
the assumed date, the most natural supposition is
that he would take the date of Full Moon from the
Calendar, and calculate, employing a rough and
generally understood average, its position on any
subsequent day by merely counting so many days
from that date. We should expect him as a matter
of fact (according to the words of ' Il Maestro ')
τὴν ἀκρίβειαν ἐπιζητεῖν . . . . κατὰ τὴν ὑποκειμένην ὕλην,
καὶ ἐφ' ὅσον οἰκεῖον τῇ μεθόδῳ . . . . ὅπως μὴ τὰ πάρεργα
τῶν ἔργων πλείω γίγνηται.

Again, though I am not (happily) very familiar
with the aspect of the heavens ' about 3 a.m.,' I do
not imagine that so early as that, ' the dawn is just
beginning (as Mr. Butler says) to whiten in the East,'
in the first half of the month of April ; nor is there, I
think, any special propriety in saying (as he suggests)
that Scorpio, which ' is just on the meridian ' at that
hour, may be said to be on the *forehead* of the dawn.
By *passi*, in l. 9, Mr. Butler understands ' signs,' but
he candidly admits that ' there is, on any explanation,
some confusion in lines 7–9.'

I ought perhaps to say a few words as to the

theory that the Solar Aurora is referred to, though I regard that idea as so utterly unsuited to the passage itself and all its surroundings as scarcely to deserve serious refutation. It will be found discussed and sufficiently refuted in *Scart.* p. 154. I will only again note, among many other objections, that in this case the term *concubina*, as applied to Aurora the *wife* of Tithonus, becomes both meaningless and offensive. (See note above on p. 84.)

Still less worthy of consideration perhaps is the curious modification of this view, that Dante refers to the Solar Aurora *in Italy*, in contrast with the nocturnal phenomena of lines 7–9 *in Purgatory*, *i.e.* 'nel luogo ov' eravamo.' It is ingeniously but fancifully argued that if it was about 2½ hours after nightfall in Purgatory, it would be 2½ hours after sunrise in Jerusalem, and consequently—Italy being, as has been already pointed out, according to Dante's geography about 45° W. Longitude, which would be equivalent to a retardation of 3 hours of time—about half-an-hour before Sunrise in Italy. No doubt *Purg.* iii. 25 and xv. 6, which we have already discussed, might be quoted as parallel cases of the contrast between the hours in Purgatory and in Italy, but it should be noticed that Dante is very careful there to indicate the contrast. Here it would surely be preposterous to suppose that all this brilliant description refers to an absent and invisible phenomenon. Further, what meaning are we to attach to lines 4–6 on this supposition?

Though I cannot here attempt an adequate examination of other views that have been held, I must not pass over without notice the extraordinary interpretation which has been suggested in recent years, first I believe by Prof. Antonelli, and expounded in his *Studi Speciali* (Florence 1871) and also in his Paper '*Sulle dottrine astronomiche della Divina Commedia,*' contained in the collection entitled '*Dante e Suo Secolo.*' This is adopted, and is defended, keenly and vigorously (as usual), by Scartazzini. It is this :—First of all, he reads *Titan* (*i. e.* the Sun) instead of *Titone* (or *Tithonus*) in l. I [1], adopting a variant which is found in the great Vatican MS., denoted as 'B' by Witte, but has not (as far as I

---

[1] In regard to the irregular form *Titone*, instead of *Titono*, from *Tithonus*, it may be remarked that old Italian abounds with every kind of anomalous interchange of the forms appropriate to different noun-declensions in Latin. A great part of Nannucci's elaborate *Teorica dei Nomi* is taken up with examples of such irregularities in every declension, '*per uniformità di cadenza.*' I select, out of many instances like the one before us, the following :—Radamante (*Boccaccio*) ; Berlinghiere (= Berengarius), Sonnolente, Turbolente (all three in *Pulci*) ; Nile and Menale, in the *Dittamondo.* Also in Dante himself, frodolente (*Inf.* xxv. 29) and pome (*Par.* xxvii. 45 and 115) Nannucci quotes one authority for *Titono. Per contra*, we very frequently meet with such forms as *Tritono, Clemento, Etiopo*, (*e. g. Purg.* xxvi. 21), *Apollino, Cesaro, Atlanto*, &c. The alleged irregularity then of the form *Titone* is no reason for preferring the more regularly-formed *Titan*. These anomalies would be facilitated by such truncated forms as *Titon, padron, Nin* (*Purg.* viii. 53), *Nil* (*Par.* vi. 66), &c.

know) much other authority. The main outlines of the interpretation are then as follows:—(1) that neither Solar nor Lunar Aurora are referred to, but that 'La concubina di *Titan*' represents Tethys, the wife of Oceanus, and is in fact an equivalent to *onda marina.* The following lines are therefore tantamount to saying that the ocean waves towards the East were illuminated by light, probably from the rising Moon, but at any rate from some source *other than the Sun* (for thus Scartazzini strangely interprets l. 3).

(2) He thinks that the *freddo animale* is not the Scorpion, or indeed any other Constellation whatever, but only certain stars, among which might be some of those of Scorpio, '*disposte in forma di serpe,*' the serpent being well known as 'frigidus anguis.'

(3) He interprets *passi* in l. 7 of the hours of the night, as we have done above, and consequently his final conclusion is precisely that which we have come to, *viz.* that the hour was towards 9 p. m. 'Μεταβαίνων δὴ ὁ λόγος εἰς ταὐτὸν ἀφῖκται.'

Now I must say that this interpretation involves such a congeries of improbabilities or difficulties in every single line, that I think for once Scartazzini's usual judgment and common-sense seem to have strangely deserted him. I take his interpretation line by line.

(1) In *l.* 1, Scartazzini (as we have seen) censures Dante for describing the Lunar Aurora as the mistress

of Tithonus, as thereby *falsifying* (his word is *falsificare*, p. 152) Mythology. But what are we to say about his own theory that Dante here instals the Ocean Wave as the mistress of Titan or the Sun? Where is there any trace of this notion in Ancient Mythology? He quotes, it is true, a number of classical passages from which such an unpleasant inference *might* conceivably have been drawn. But there is not the least evidence that it ever *was* drawn, and it involves quite as much of an *adaptation* or *falsification* (whichever you will) of traditional mythology as that of describing the Lunar Aurora as the mistress of Tithonus.

In short all the rhetorical and prudish nonsense on p. 152 as to this latter idea involving 'lordura,' and its being 'sozza pittura' from which we are compelled 'svolgere con nausea e con ribrezzo gli occhi' (!), all this I say applies with at least equal force to his own contribution to mythology as just expounded. Indeed I should maintain that these expressions apply with much greater force against his invention that Tethys is the mistress of the Sun, since she is undoubtedly the lawful wife of Oceanus, whereas Dante's 'Concubina di Titone antico' *i. e.* the Lunar Aurora, has at any rate no other known attachment.

(2) Next in *l.* 2, the word *balco* (*i. e.* gallery or balcony) implies some elevation, and clearly indicates some phenomenon *in the sky*, not on the 'suol marino,' as Dante calls it. It loses its significance then when

applied to light on the waves, though it is appropriate to the light of dawn on and above the horizon. *Scart.* paraphrases *balco* by *lembo*, which is clearly inadequate [1].

(3) In *l.* 3, *Scart.* takes *fuor delle braccia* apparently to mean that the wave, *i. e.* Tethys or the *concubina* of the Sun, was illuminated *otherwise than by the Sun* her 'dolce amico.' This is surely an impossible, or extremely forced meaning for the words. Moreover, considerations of language apart, it can scarcely be doubted that *fuor delle braccia* represents the idea of

$$\text{ἐκ λέχεων παρ' ἀγαυοῦ Τιθώνοιο}$$

in the well-known passage of Homer (*Il.* xi. 1) or again the *Tithoni linquens cubile* of Virgil (*Geor.* i. 447).

---

[1] I have no doubt that *balco* (as indeed *Scart.* reads) and not *balzo* is correct here. See Blanc's *Dizionario s. v.* Balzo. The latter word occurs several times in the Divina Commedia (including twice in this Canto, lines 50 and 68); the former here only. An alteration then to the *lectio facilior* would not be improbable. But apart from this, the change would be so slight that it might have arisen accidentally. Every one familiar with MSS. is aware that *z* is very commonly written *ç*; the difference between the two words therefore would only be that between *c* and *ç*. I observe that in my own MS. of the *Divina Commedia*, the cedilla has been in fact added to the original *c* by a later hand and in different ink. The word *balco* is more appropriate, since *balzo*, wherever Dante uses it, seems to mean a rocky projection, but as it also in all these places involves the idea of some elevation (see especially *Inf.* xi. 115; xxix. 95; *Purg.* iv. 47; vii. 88), either reading would equally sustain the argument in the text. Blanc says of *balzo*, ' sembra indicare propriamente uno sporto, un terrazzino, o roccie sporgenti.' Finally, Buti's explanation of *balzo* or *balco* (whichever he read) is ' è luogo alto dove si monta e scende.'

(4) In *l.* 4, his explanation takes no notice of the expression *fronte* as applied to the Ocean wave. What is the point of it in that case? On the other hand, the 'front' or forehead of the dawn speaks for itself.

(5) In *lines* 5 *and* 6 he maintains that '*freddo animale*' has nothing to do with Scorpio (his objections being very feeble and unconvincing), but only indicates some stars arranged in the form of a serpent. To which I would object (1) What are these stars, and are any such so well known as to convey any definite idea to a reader of the passage? (2) Why should it be worth while to mention that there were stars in the neighbourhood arranged in this form? (3) The extended description *freddo animale Che con la coda percuote la gente,* if it is only to say that the stars were arranged in a serpentine form, is rather superfluous, and, what is more important, it is much less appropriate (not *more*, as Scartazzini strangely asserts) to an ordinary serpent than to a Scorpion. The (as I should say) obvious imitation of Ovid already pointed out, not only shews the appropriateness of such a description as applied to the Scorpion, but seems almost to settle the question that it is to be thus understood.

Finally, let us ask ourselves what is the purpose for which this reference to time is here introduced, and what consequences in any case flow from the interpretation adopted? It describes the hour when

Dante sunk down overpowered with sleep; and there is no difficulty, as far as I can see, in supposing a long sleep of as much as ten or eleven hours, especially, as *Scart.* suggests (whose interpretation, like our own, would involve this supposition), after the extreme fatigue of the last three nights (p. 155). How oppressive that fatigue was Dante frequently reminds us, and in fact gives us himself a hint here in the phrase 'che meco avea quel di Adamo.' Observe also that the last reference was to twilight (*viz.* in viii. 49) and that there is nothing recorded meanwhile to require much lapse of time; further, that the incident noticed still nearer to our present point, *viz.* in l. 95, the approach of the snake which the heavenly watchers detected, seems naturally to coincide with nightfall [1]. Those who argue for the Solar Aurora lay stress on the undue length of time allotted to this sleep, and the alleged insufficient time on the rival theory for the preceding interviews. *Per contra*, the advocates of the Lunar theory object that the interval allowed by their opponents for sleep is too short (*e.g. Scart.*, p. 155). Here however we are in the region not of conclusive arguments, but mere conjectures of probability, and whether Dante's imaginary sleep is more likely to have been repre-

---

[1] Indeed we have a direct intimation that it was so associated in l. 39, where the advent of the two guardian angels at dusk is explained to be because of the expected *imminent* approach of the serpent, 'che verrà via via.'

sented as lasting ten or eleven hours or only five or
six, cannot be positively determined.

There is however an argument on this subject of
the probable duration of Dante's sleep which is
employed by Mr. Butler, (one of the advocates of
the Solar Aurora,) which seems to me to point
quite the other way. He says (p. 111) that 'the
analogy of the two following nights would make it
probable that Dante does not fall asleep till towards
morning.' But, apart from the consideration that in
Purgatory, as the region of light and hope, Dante
would be likely to associate as much as possible of
his journey with daylight, the actual comparison suggested by Mr. Butler seems to me to point in the
same direction. Turning to xxvii. 70, we find that
on the next night but one (*i.e.* April 12) sleep overpowered him even earlier than the hour of nine here
advocated. Very soon after sunset, of which they
became aware—note the graphic touch—by the cessa-
tion of the shadows as they were journeying East-
wards, and therefore with their backs to the Sun
(see lines 68–9), and, as Dante expressly says, *before* the
vast horizon had yet assumed one uniform dark hue,
and *before* night had fully distributed herself, he at
any rate lay down to sleep ('fece letto,' l. 73; see
also lines 88–93) guarded by his two companions as
a goat by its shepherds (l. 76, etc.) *i.e.* distinctly *not
later than* 8 *p.m.*[1] The case of the night next succeed-

---

[1] In fact he lays stress more than once on the fact that onward pro-

ing the night here referred to in Canto ix, is indeed
different, but the difference is very significant, and if
there ever were an 'exception that proves the rule'
it is surely this. It is towards midnight (see xviii. 76),
and Dante is heavy with sleep (l. 86) :—

> Stava com' uom che sonnolento vana.

But this drowsiness is suddenly dispelled, and how?
By a crowd of spirits who came running by, extolling
examples of activity and energy as a counterblast to
the sin of *accidia* or spiritual sloth, which is the sin
specially expiated on that *Cornice* of Purgatory (see
esp. lines 88–98, 115–118, 127–8). Thus on the night in
question, though progress is as usual suspended after
dark, the discourses are continued and companies of
spirits accosted and conversed with till nearly midnight.

Is not this subtle touch, I mean the delay even of
the natural claims of sleep in this place as a protest
against *accidia*, thoroughly Dantesque? And may we
not then say that the analogy of the other two nights

gress after nightfall is impossible in Purgatory (in evident allusion to
such passages as S. John ix. 4, &c.). See in illustration of this, *Purg.*
vii. 44, 49–57 ; xvii. 61–3; xviii. 110; xxvii. 75, &c. It is true he does
not definitely say when he actually fell asleep on the night in question
(l. 91), and at any rate he did not do so till he had had time to observe
the unusual brightness of the stars (l. 89), but that need not have been
long ; and certainly the graphic picture of the goat guarded by its
shepherds loses much of its force unless the goat were asleep. Moreover we can scarcely suppose that Dante and his two companions lay
awake and yet silent for several hours. (Contrast *Purg.* xvii. 84.)

in Purgatory is distinctly in favour of the sleep in Canto ix. having been a long one, and covering the greater part of the hours of darkness?

IX. 13, 52. At length to pass on, we find in this Canto, l. 13, and again l. 52, that this sleep, like that of each night in Purgatory, is terminated by a dream, and in each case towards the hour of dawn, that hour when, as Dante reminds us here and elsewhere, following the general tradition of classical antiquity, the visions of dreams are truthful and prophetic. Compare especially *Inf.* xxvi. 7

> Se presso al mattin del ver si sogna,

or as Ovid (quoted *l. c.* by Scartazzini) says,

> Tempore quo cerni somnia vera solent.

On this night he has the vision of the eagle here described, on the following night that of the Siren (see xix. 1–34); and on the third night that of Leah (see xxvii. 94–108).

IX. 44. In l. 44 he describes his awakening when the Sun had already been up more than two hours, so that the time would be about 7.30 a.m. Virgil then expounds his dream, shewing how it represented his transportation during his sleep by the aid of Lucia to the actual gate of Purgatory itself. Hitherto, it will be remembered, they have been traversing the Ante-Purgatory only.

X. 14. The next reference is in x. 14, where the

Moon-setting of April 11 is referred to. I should have calculated this at about 8.15 a.m. for the Calendar, and 10 a.m. for the Real Moon, on the same principles as before. I observe that Philalethes puts both about an hour later. Also that Scartazzini makes the calculation that this being 4½ days (so he says) after the *plenilunium* of Thursday night, there would be a retardation of the Moon behind the Sun of about 3 hrs. 54 min. I can only count 3½ days which would give a retardation of about three hours, but I suppose Scartazzini (with whom I find since Della Valle agrees) must take this to be the morning of Easter *Tuesday*, and not as we have contended, Easter *Monday*. As then the Sun rises about 5.15 at that time, we should arrive, as I should say, at about 8.15, or as the authors just quoted would say 9.15, for the approximate time of Moon-setting. Now Dante awoke when the Sun had been up fully two hours, *i. e.* say about 7.30. Observe that this datum as to the Moon's position is not very definite, as he merely says that it *had* already set; *i. e.* they found that it had *already disappeared* (but it is not said how long) when they emerged from the narrow rocky passage leading from the Gate of Purgatory to the 1st *Cornice*. Thus we should have, according to the various calculations, about one, or two, or three hours occupied in passing it. One hour on my supposition ; two by the calculations of Scartazzini, Philalethes, &c.; and three or more, if the Astronomica Moon be regarded. This however, being an arbitrary and imaginary period, gives us no certain datum for argument as between the Calendar and Real Moon, but on the assumption that the day is March 29th, the allusion would be wholly meaningless, as the Moon then set about 9.30 *p.m.*, about as far wrong as it could possibly be.

XII. 81. In xii. 81, Midday has just passed as they are leaving the 1st *Cornice* of Purgatory, where Pride is punished, and the Angel of the 6th hour of day, returning from her service, points them out the way. (Compare the language used in xxii. 118.)

XV. 1. In xv. 1 (a passage which we have already discussed), we have the hour of 3 in the afternoon indicated as they pass from the 2nd *Cornice* (that of Envy) to the 3rd, which is that of Anger. In l. 141 the Sun's rays were low and full in front, *i.e.* they were still 'stepping westward,' as we learn from the expression 'contra i raggi,' &c. in l. 140, and the rather quaint and blunt phrase, that the rays smote them 'per mezzo il naso' in l. 7 of this same Canto. We may observe that, besides the discourses recorded since the last note of time, Dante had wandered fully half a league in a state of semi-conscious ecstasy (see l. 121, &c.).

XVII. 9. In xvii. 9, as they are leaving the 3rd *Cornice*, the Sun is on the point of setting, and in the lower valleys his light had already departed (see l. 12). As they ascend to the 4th *Cornice* where *Accidia*, or

Sloth, is punished, twilight has come on, the last light in the sky is rapidly fading, and the stars are beginning to appear here and there (see lines 62 and 70–2).

XVIII. 76. We now come to another passage of considerable difficulty, regarded as an indication of time, *viz.* xviii. 76.

> La luna quasi a mezza notte tarda
> Facea la stelle a noi parer più rade
> Fatta come un secchione che tutt' arda.

The majority of Commentators have assumed (as it appears to me quite needlessly), that this must refer to the actual hour of Moon-*rise*, which would certainly be, according to the principle we have been advocating, about 10 p.m. or perhaps 10.30, since the Moon is already well up, and producing a sensible effect in quenching the lesser stars. Hence however, grave difficulties have been raised by Della Valle, Ponta, and others (see Scartazzini, note *h. l.*), and they have endeavoured to edge back the *plenilunium* about twenty-four hours by a somewhat forced method of reckoning, so as to gain an hour or so nearer midnight for the Moon-rise on this night [1]. But even if Moon-rise were undoubtedly indicated here, we might,

---

[1] Della Valle says that the Full Moon should be taken as occurring about 1 or 2 a.m. on April 7th, *i. e.* practically soon after midnight on what we should popularly call Wednesday night, though no doubt, astronomically speaking, Thursday morning ; but then Dante could not have described this without misleading as 'iernotte,' when speaking on Saturday.

I think, maintain that the approximation pointedly
indicated by *quasi* might cover the interval even of
nearly 2 hours. Such in fact seems to be the view
of Philalethes, who says the Moon rose ' *Etwa um*
10 *Uhr, also schon ziemlich gegen Mitternacht.*' But
though I do not consider it necessary, merely ' θέσιν
διαφυλάττων,' to explain the passage otherwise, I do
not think it at all certain that Dante intends to speak
of the actual hour of Moon-*rise* at all. I would suggest
—(1) it is certainly not distinctly indicated, or necessarily implied, by his words.   For instance, I
observe that, among others, Mr. Butler explains the
passage otherwise, *viz.* of the Moon's 'southing,' and
he even maintains that 2 or 3 a.m. is intended, which
however I cannot agree with.   (2) The effect here
indicated of the quenching of the lesser stars by the
light of the gibbous or pitcher-shaped moon (as it is
graphically described in l. 78) would be much more
striking if it were some little time above the horizon
than if it were just rising.   (3) I think it probable
that the whole passage is only a poetical and slightly
elaborate way of saying that the *hour* was approaching
midnight, described, as usual, by some striking visible
aspect of the fact.   It is not half so elaborate or
artificial a way of describing a simple fact or phenomenon as many other passages that might be cited.
It is surely quite a natural (poetical) description of
such an hour (it being allowed that the Moon was up
as a fact), to say, ' And now the Moon, as it were

towards midnight late, shaped like a pitcher all afire, was making the stars appear to us more rare.' Lubin is in no difficulty here, finding that his 'Real' Moon rose about 11.50 p.m. I see that Antonelli (*ap.* Scartazzini) substantially interprets the passage as I have suggested above, though I do not think it at all necessary to adopt his explanation that *tarda* agrees with *notte* and not *luna* ('Quasi alla tarda ora della mezza notte la luna . . . faceva,' &c.). It would surely be a very common poetical device to transfer the epithet from the night itself to the Moon or any other object thus 'lated in the night.'

In any case I would observe, that if this does refer to Moonrise, and if we must also assume (as some Commentators do) that *therefore* that phenomenon took place about 11.45 p.m. (as would be the case with the 'Real' Moon), or in any case literally near midnight, *then* we get the utterly impossible hour of between 10 and 11 p.m. *for the Moon-rise of the previous evening,* in the celebrated passage at the beginning of Canto ix, assuming, as I think we must certainly assume, that the *Lunar* Aurora is there intended. I say the hour 10 to 11 for the Lunar Aurora on the previous night is *impossible,* as it would not suit *passi* in l. 9, and the Right Ascension of the Moon at that age would bring her some way behind the Scorpion into Sagittarius. If then 9.15 (or thereabouts) be absolutely required for Moon-rise on the *previous* night, we cannot possibly get beyond 10.15

or 10.30 for Moonrise (or some time shortly after), on
this night [1]. So we are driven absolutely to choose
between these two alternatives—*either* (1) of supposing
a liberal margin to be covered by *quasi* ; *or* (2) of supposing the passage not to refer to Moon-*rise* at all ;
either alternative being I think quite admissible.

The words which follow in l. 79 describe evidently
the backing of the moon through the signs from
west to east [2], which causes the daily retardation
to which we have so often referred : and more par-
ticularly he says that she was in that path of the
Zodiac which is illuminated by the Sun, when the
people of Rome see him setting between Sardinia
and Corsica. This is stated by Mr. Butler, no doubt
correctly, to be towards the end of November, when

---

[1] There is a curious variant which I find in Buti only of the old
Commentators (though Witte notes it as occurring in the margin of
the *Santa Croce* MS.), *viz.* 'a *terza* notte,' instead of 'a *mezza* notte,'
which Buti explains, 'quasi passata la terza parte della notte.' The
expression would surely be a most unusual one, and it looks like an
emendation made by some one who had worked out the calculation,
and thought the discrepancy between 10 or 10.15 p.m. and midnight
too large to be covered by 'quasi.' I was not aware of the existence
of this variant in time to get collations of the passage. I find it
recorded however in different quarters as existing in several MSS.,
including some of considerable celebrity, especially *Cod. Fil.* at Naples
(Batines, No. 411); *Cod. Cavriani*, at Mantua (Batines, No. 244);
Cod. Gambalunga, at Rimini (Batines, No. 404); also those num-
bered by Batines 11, 124, 147, 249, 405.

[2] Similarly in *Par.* ix. 85 the phrase *contra il sole* is used to indicate
eastward direction ; and again in *Par.* vi. 2, the removal of Constan-
tine of the seat of Empire from Rome to Constantinople is described as
contro il corso del ciel.

the Sun sets west by south. If so the Sun would be then in Sagittarius, and that is precisely where the Moon's Right Ascension would bring her on this night, as is pointed out by Della Valle. (See Scartazzini's note.) Dante's indication of the Sun's position here, as seen from Rome, is curious. These islands being invisible from Rome, the Sun can only be said to be seen setting between them, from a knowledge of their position on the map compared with the observed direction of the Sun [1]. In this sense only can it be true that (as some of the old Commentators say), Dante observed this himself when at Rome; and in this sense it is very likely indeed to have been true, since he was actually at Rome at the moment of the disastrous entry of Charles of Valois into Florence on November 1st, 1301, and for some time afterwards, *i. e.* at the very time of year here described.

But I must hasten on, for

> Lo tempo è poco omai che n' è concesso
> Ed altro è da veder che tu non vedi.
> <div align="right">(<i>Inf.</i> xxix. 11, 12.)</div>

though fortunately no very serious difficulty is presented by the remaining passages.

XIX. 1–6. In this passage we have the hour before dawn on Tuesday, April 12th, described by two

---

[1] Compare the statement as to the Moon setting beneath Seville in *Inf.* xx. 126.

indications. (1) It was the coldest hour of the twenty-four; (2) the later stars of Aquarius and the foremost ones of Pisces were on the horizon. This perhaps we may be allowed here to take for granted is the meaning of the 'maggior fortuna' of the wizards (1. 4). It was a peculiar arrangement of dots, corresponding to one that can be formed out of certain stars on the confines of these two constellations. These were now in the east before the dawn (1. 5).

XIX. 37–9. In lines 37–9 it was now full daylight, with the Sun on their backs, so that they were still journeying towards the west, when they enter the 5th *Cornice*, where Avarice and Prodigality are punished. (Note how Scartazzini still doubtfully gives three alternative suppositions as to the day intended here, *viz.* March 29, or April 9, or April 12.)

Observe here also the admirable fitness with which Dante times his progress so that the time spent in the *Cornice* where *Accidia*, or Spiritual Sloth, is punished is exactly coincident with the hours of night— 'the night when no man can work.' He enters it as darkness comes on (see xvii. 70–80), and leaves it the next morning, as soon as he awakes with the 'nuovo Sol' (xix. 38), being mildly chided by Virgil for the length of his slumbers (xix. 34)[1]. I might

[1] As another passing illustration of Dante's constant regard for the 'fitness of things,' note how on the 5th *Cornice* the examples of the

perhaps mention here that it will be found that in each of the other *Cornici* he spends from three to five hours.

XXII. 118. We next find them ascending to the 6th *Cornice* (that of Gluttony), and as they approach the mysterious tree (about which so much has been written, and such strange mistakes made, especially as to the meaning of xxii. 133–4), the hour is indicated as 11 in the morning (see lines 118–9) in language resembling that in which Noon was described in xii. 81.

XXV. 1–3. The next indication of the hour that we meet with is at xxv. 1–3, while they are still on the same 6th *Cornice*, and there seems to be no reasonable doubt that about 2 p.m. is intended. This is one of the passages on which I think some superfluous astronomical ingenuity has been expended, the point being whether we are to make allowance for the retrocession of the Equinox and the error in the Calendar, and so take the Sun's true astronomical position, or whether we are to be guided by the ordinary popular notion that the Sun is in Aries for a month from March 21st onwards. The difference of the result is absolutely immaterial, as it is only

Virtue of Liberality are proclaimed *by day*—
    Quando il dì dura: ma quand' e' s' annotta,
    Contrario suon prendemo in quella vece:
and then *by night* the instances of the *Vice* of Avarice are denounced (*Purg.* xx. 101, &c). In l. 121 this distinction is again emphasized that we may be sure not to miss it,
    Però al ben che *il dì* ci si ragiona.

a question between about 12.30 and 2 p.m., either hour here being quite arbitrary and fictitious. Here again however I think it is more probable that Dante adopts the sense in which ordinary people would be most likely to understand his words, just as we popularly refer to the indications of the compass as it stands, without allowing for the magnetic variation, though we are quite aware that it amounts in England to a no less serious difference than about 23 degrees. If this be the way to interpret the passage, the Sun being now rather backward in Aries, the time when Taurus is on the Meridian of Noon, and the opposite sign of Scorpio on that of midnight as here described, would be generally understood to be about 2 p.m. though, as each constellation covers many degrees of space, the indication is only an approximate one.

XXVI. 4–6. In xxiv. 4–6 they are on the 7th and last *Cornice*, where Lust is punished, and the time is apparently about 4 or 5 p.m., since the Sun is getting low in the west. This is indicated by two circumstances, (1) the blue of the western sky is turned pale by his light; and (2) his rays strike them on the shoulder, which indicates a low altitude.

Note the continuation of this magnificent passage. Dante's body does not cast a shadow here as so often elsewhere in his passage through Purgatory, falling as it does on the burning flame in which these spirits are being purified. The only effect produced by it

is that the flames appear more ruddy where it inter-
cepts the Sun's rays. Observe the words '*pure* a
tanto indizio' (l. 8). Some of my readers may
remember that these few lines are quoted by Mr.
Ruskin (*Mod. Painters*, ii. p. 259) as probably the
finest description in literature of intense heat. He
maintains that in these few very simple, and in
some sense commonplace, touches, Dante, 'with no
help from smoke or cinders,' has produced a more
vivid effect than Milton has secured in ten lines of
elaborate description and varied imagery. Dante's
few words suggest, as Ruskin says, 'lambent annihi-
lation[1].' I wish I had space to illustrate further
this splendid and unequalled power in Dante of
piercing at once to the very heart of things, and
revealing as it were a whole world of scenery, or
of emotion, or of passion, *at a flash*; and as often
as not by a flash of silence, that is more eloquent
than any words.

XXVII. 1–5. In this passage (which we have
already explained), the hour is approaching Sunset,
'il giorno sen giva,' though, as we learn from lines
61–3, the Sun had not yet disappeared. This is again
one of the passages on which a good deal of perverse
ingenuity has been spent, some Commentators calling
in the aid of refraction to show that by the time of
Sunrise at Jerusalem the disappearance of the visible

---

[1] We might perhaps quote the terzina, *Purg.* xxvii. 49–51, as another
striking passage of the same kind.

Sun would not yet have occurred at its Antipodes, the Mountain of Purgatory, for seven minutes later! This is surely rather 'learned trifling' with the language of poetry.

XXVII. 61-9 ; 70-2 ; 89, 90. In lines 61–9 we have the actual going down of the Sun; the gradual coming on of darkness in lines 70–72; and in lines 89–90 the shining out of the stars 'clearer and larger than their wont.' This brings us to the end of the third day, Tuesday, April 12th, and the poets have now reached the end of Purgatory proper. The dawn of the fourth day is beautifully described in lines 109, &c.; the Earthly Paradise is entered, and Virgil takes his leave in the splendid passage with which this Canto ends, in the course of which he points to the now fully risen Sun:—

   Vedi là 'l Sol, che in fronte ti riluce (l. 133).

XXXIII. 103. Our last reference is in *Purg.* xxxiii. 103, where, after the mystic procession and the Vision prophetic of the disasters impending over the Church have passed away, at the hour of Midday [1], Dante is

---

[1] Here again (l. 103) Antonelli has indulged in much subtlety of physical and astronomical speculation as to the causes of the greater brightness and slower motion attributed to the Sun. I have no doubt the simple explanation suffices that both these are obvious characteristics of the Sun at that hour and position, especially when I find Dante recurring to the latter phenomenon at any rate in order to describe the southern quarter of the sky in *Par.* xxiii. 11, 12,

   Rivolta inver la plaga
  Sotto la quale il Sol mostra men fretta.

led by Matilda, at the command of Beatrice, to drink of the water of Eunoe, and he is then prepared to ascend to Heaven:

> Puro e disposto a salire alle stelle.

And now to sum up in rapid review the results of this long investigation.

(1) That the year of the Vision is to be taken as 1300, rather than 1301, is I think beyond question. I do not pretend to have discussed that subject at all exhaustively, nor do I claim to have settled the question by independent arguments, which our present limits would have made impossible. I have rather appealed to the united and unhesitating declaration of all the principal writers or Commentators on the Divina Commedia in various ages, that no reasonable doubt whatever exists that 1300 is the true date, as almost absolving us from the necessity of re-opening the question.

(2) Next as to the date within that year, I think it will at any rate be agreed that the supposition of Dante having adopted any other date than the actual Good Friday and Easter of 1300, and especially the notion that he refers to March 25th, &c. may be confidently dismissed.

(3) As to the view for which I have been contending, *viz.* that he assumed the Full Moon to be on April 7th, *according to the Calendar*, and not on April 5th, as it would be according to strict astronomical cal-

culations, I venture to think that this suggestion, however simple it appears, affords a key by the help of which we obtain a consistent and satisfactory interpretation of the various time-references depending on the Moon's position throughout the poem. I do not myself see (though of course I may very likely be mistaken) that it leaves any serious difficulty unexplained, while, even apart from this, it seems to be a more natural supposition than any other. Indeed, I very much doubt whether Dante was even aware of the discrepancy between the two Moons, or, if he were, whether he had access to any means for correcting it.

(4) We note the extraordinary vividness with which Dante, writing some years later, kept constantly before his eyes all the circumstances and details appropriate to the assumed date 1300. He seems never to have forgotten this. I at least know of no single case in which he has spoken of any event as past (even though, like the death of his friend Guido Cavalcanti it occurred very shortly afterwards, and perhaps even late in the year 1300 itself) which had not actually happened at the assumed date of the Vision. It seems not improbable that an event even so soon afterward as May 1st, 1300, is referred to under the form of prophecy[1]. (Some illustrations of this will be found in a supplementary note.) Surely never did any writer in this, and in all other respects,

---

[1] I refer to the prophecy in *Inf.* vi. 64-5.

more accurately fulfil the precept of Aristotle in the
Poetics: — Δεῖ τοὺς μύθους συνεστάναι καὶ τῇ λέξει
συναπεργάζεσθαι ὅτι μάλιστα πρὸ ὀμμάτων τιθέμενον· οὕτω
γὰρ ἂν . . . ὥσπερ παρ' αὐτοῖς γιγνόμενος τοῖς πρατ-
τόμενοις, εὑρίσκοι τὸ πρέπον καὶ ἥκιστα ἂν λανθάνοι τὰ
ὑπεναντία. (*Poet.* xvii. *init.*)

The passages here discussed have been taken, as
will no doubt have been observed, entirely from the
Inferno and *Purgatorio.* It is hardly necessary to add
that Dante gives us no such marks of time in the
Paradiso [1], since there he has passed from time to
eternity,—

> All' eterno dal tempo era venuto.
>
> (*Par.* xxxi. 38.)

Also there they have 'no need of the Sun, neither
of the Moon, to shine in it,' for 'there is no night
there.'

---

[1] See Supplementary Note vi, p. 126.

SUPPLEMENTARY NOTES.

[The following notes are printed here, as some of them seemed
too long to appear as foot-notes on the pages to which they
severally refer, and others relate to works that have only
been met with since this Essay was written.]

I. *On the selection of the year* 1300 *for the Vision.*

[As an additional note on p. 8.]

DEAN PLUMPTRE in the valuable Introduction to his
Translation recently published (pp. lxv–lxvii) makes an
interesting, and I think very plausible, suggestion that the
assumed date of the Vision had a very special significance
for Dante himself as the turning-point in his own life (see
especially *Purg.* xxx. 130-8). We know from his own
confessions in the *Convito* and elsewhere, how, after the
death of Beatrice, he 'forsook his first faith,' and sought for
consolation from the 'broken cisterns' of heathen philosophy,
and from

le presenti cose
Col falso lor piacer. (*Purg.* xxxi. 34.)

Moreover, while asserting as a fact in his own history his
recovery from these errors, and his return to Faith and to
his 'first love,' he also distinctly connects that recovery,
both in time and cause, with the contemplation of the
scenes and subjects of this Vision. Dean Plumptre suggests
that Dante may have been actually at Rome, and perhaps
for the first time in his life, at the date assigned to his

Vision, *viz.* Easter-tide of the year 1300 ; and that this was in fact the turning-point in his life from speculative, and more or less sceptical philosophy to Revealed Truth, as symbolized by Beatrice. If so, the assumed date was not an arbitrary poetic fiction, but corresponded to a fact of the deepest interest in his own history, 'the conversion crisis' of his life (p. lxxvii). This would satisfactorily account for the pointed and repeated references by which the date assigned to the Vision is emphasized. It is fixed, as the Dean says, 'with a precision the only natural explanation of which is that it represents *a fact.*' At any rate this suggestion is a highly interesting one, though it cannot be regarded as more than an ingenious conjecture. It must not be forgotten however that in *Inf.* xviii. 28–33 we have very strong evidence that Dante was at Rome *sometime* in the year 1300, and the *esercito molto* would naturally be at its height at the solemn season of Holy Week and Easter.

II. *On the assumed date* 1300 *never forgotten by Dante.*

[As a note on pp. 8, 136.]

The following are, as far as I can recollect, the chief passages bearing on this. I will mention first those which are sometimes thought to be exceptions, or oversights, on the part of the poet.

I. *Inf.* xviii. 28. Dante here refers to an incident occurring at Rome during the progress of the Jubilee in 1300. This was proclaimed by Boniface on Feb. 22 in that year. But (1) Dante may have been there in March or at Easter itself. (See last note.) Certain it is that early in 1300 two embassies went to Boniface from Florence, and as Dante took part in such an embassy shortly after, he may have accompanied either of these. (2) Whether he was there or not, the incident he mentions may have occurred at or before Easter in that year. (3) Whether it· did so occur or not, the comparison is introduced here not as having been spoken of *by or to himself* during his pilgrimage, but as an illustration given by the Poet *when narrating that Vision afterwards.*

(I mention this at some length because Grion uses this as an argument for the assumed date of the Vision being 1301.)

II. Precisely the same principle would apply in reference to the event referred to in *Inf.* xix. 19. Grion (p. 10) states that this occurred, according to Jacopo di Dante, on April 1st, 1301. Again I would say, this is a statement evidently made by the Poet *while writing his narrative,* and it does not profess to have been referred to by himself or any of the characters introduced by him *during the Vision itself.* I might add that this would not help Grion or any one else to maintain 1301 as against 1300, since the expression *ancor non è molt' anni* is equally (from that point of view) inapplicable to either date.

III. The '*ruina*' known as 'Slavino di Marco,' near Riva, which is supposed with much probability to be referred to in *Inf.* xii. 5, &c., is stated to have occurred on June 20th, 1309. But granting both this date and the explanation of the allusion to be beyond doubt, this also is only referred to by the Poet *as writing afterwards* by way of an illustration, and we certainly need not suppose him to have been so pedantic as to refuse an apt comparison from a natural phenomenon then familiar to his readers, because it was non-existent at the assumed date of his poetic pilgrimage.

IV. Another case deserving notice occurs in *Purg.* viii. 74, *viz.* the second marriage of Beatrice, the widow of

Nino Visconti with Galeozzo Visconti, which was celebrated 'with extraordinary pomp' (according to the chronicles) on June 24th, 1300. If so (as Scartazzini observes) she was no doubt betrothed to him for some time previously, and had 'put off her widow's weeds,' which is all that Dante commits himself to, before the assumed date of the Vision. She had become a widow, it may be added, about four years previously, in 1296.

V. *Purg.* xiii. 152 is supposed to refer to the purchase of Talamone by the Sienese, on Sept. 10th, 1303, for 8000 golden florins, during the time (as it is stated) when Dante was himself at Siena (see Aquarone, *Dante in Siena*, p. 70). No doubt if he were then present it would account for the impression which the incident made upon his mind, and it may have suggested an allusion to it. But his statement is far from being so definite as is assumed. Dante only says that the hearts and hopes of the foolish Sienese were set upon the place, and this may well have been the case for three or four years before they succeeded in securing it.

Per contra, we have some remarkable instances of events occurring *very soon* after the assumed date which are spoken of as still future, and referred to under the guise of prophecy. The following are a few cases which occur to me :—

Inf. vi. 64. The bloodshed of May 1st, 1300, at Florence (probably). (See Giov. Villani, *Chron.* VIII. c. 38.)

Inf. x. 111. The death of Dante's friend Guido Cavalcanti in the winter of 1300–1.

Purg. xiv. 118–9. The death of Mainardo Pagano, 'il Demonio,' Aug. 16, 1302.

Purg. xviii. 121. The death of Alberto della Scala Sept. 10th, 1301.

III. *On the date of the Vernal Equinox, and traditions of the Creation.*

[As a note on pp. 14–16.]

Brunetto Latini seems to have placed the Vernal Equinox on March 18th and the Summer and Winter Solstices on June 17th (*all' XV dì all' uscita del mese di giugno*, Tes. ii. 43, part 2) and December 17th respectively, but, as far as I can see, gives no reason for this position of the Equinox beyond that '*dicono molti savi che 'l fu XIIII dì all' uscita del mese di Marzo*' (Tes. I. c. 6), or, as it stands in the original, '*XIIII jors a l'issue dou mois de Mars.*' There are, as far as I know, two other passages besides that just cited when a similar statement occurs :—*Et sappiate che 'l primo dì del secolo entrò el Sole ne lo primo segno, cioè in Ariete. Et ciò fu XIIII dì all' uscita di Marzo*, et altresì fa egli ancora (II. c. 42). And again in II. c. 43, part 2, the same statement is made in nearly identical words, and in the same chapter the Autumnal Equinox is similarly put *a XV dì all' uscita di settembre*, i.e. Sept. 18th. The words *et altresì fa egli ancora* seem to make it doubtful whether Brunetto was aware of the error in the Calendar or of the Precession of the Equinoxes, as they imply that no change had occurred in this date since the Creation. There is another passage, *Tes.* II. c. 48, where Brunetto says that all the heavenly bodies were created on March 21st ('*XI dì all' uscita di Marzo*'), that being the fourth day of the week of Creation, which he supposed to have commenced, as we have seen, on March 18th. Hence (he proceeds) many say that the Equinox is on that day. When therefore he himself puts it on March 18th, it is probably not on astronomical grounds, but because, regarding that day as '*il primo dì del secolo*,' he thought it appropriate that it should fall '*in quel buono e dritto punto,'* when day and night were equal. This view is however much more ancient than Brunetto Latini. It is found in Bede, who says that the heavenly bodies were all created 'quarta die . . . quae nunc, quantum aequinoctii conjectura colligimus, XII Kal. Apr. vocatur.' Hence the *'prima saeculi dies'* would be March 18th. So also in the curious 10th Century Anglo-Saxon Manual of Astronomy translated and edited by Wright (1841), p. 4, we read, 'The first day of this world we may find by the day of the Vernal Equinox, because the day of the Equinox is the fourth day of the Creation of this world. . . . . We will now say briefly that the first day of this world is reckoned on the day which we call XV Kal. Apr.' *i.e.* March 18th. (He quotes Bede as his authority for this.) He says later that some have placed the Equinox on March 25th, but that it is certainly ' on S. Benedict's day,' *i.e.* March 21st. So also Hippolytus, a disciple of S. Irenaeus, and apparently the first author of a Paschal Cycle, placed the Equinox on March 18th (Hefele, *Councils*, p. 318).

It is worth while adding a word as to the curious way of computing the day of the month by counting backwards, which is found in the above passages and elsewhere in the Tesoro, and in some other old writers. The month was divided in a way curiously resembling the Greek μηνὸς ἱσταμένου and μηνὸς φθίνοντος (though there seems to have been no third division corresponding to the μηνὸς μεσοῦντος), and, as with the Greeks generally, those of the former division were counted forwards and those of the latter backwards: *e.g.* Brunetto describes S. Barnabas' day as *XI dì all' entrata di giugno* (II. c. 20), and S. Thomas's day as *XI dì all' uscita di decembre* (II. c. 13). It is curious also to note that in the passage above cited from *Tes.* I. 6, the old editions, *e.g.* Ven. 1533, read ' *che 'l fu XIIII dì del*

mese di Marzo.' This is a mere Editor's correction, made
no doubt by some one who knew that March 14th more
nearly corresponded with the actual fact. It is however
unsupported by the MSS. and clearly inconsistent with the
author's repeated statements in B. II, as quoted above.

As to the common belief that the Sun was created at the
Vernal Equinox, which is of course implied in *Inferno*
i. 38–40, Dante himself gives us a curious *a priori* reason
in the very obscure passage, *Par.* i. 37, &c. and also in
Conv. II. 4 (l. 50, &c.), which passages are both noticed by
Dionisi, *Anedd.* IV. p. 51, &c. The same writer (p. 66)
also quotes a curious fragment from the Acts of an early
Council in Palaestine A.D. 196, which professed to determine '*quo modo in principio factus fuerit mundus,
id est die Domenico* (!), *Verno Tempore, in Equinoctio quod
est Octavo Kalendarum Aprilium* (*i.e.* March 25th), *Luna
Plena.'* So that thus the Creation, Incarnation, and Crucifixion would all occur on March 25th. By Creation,
when thus spoken of as the act of a single day, no doubt
the Creation of man is generally intended, though I do
not think those who indulged in these *a priori* specu-
lations were always careful to attach a precise meaning to
their words. It should be remembered also that at the time
of the very ancient Council just cited the Equinox was
held, according to the chronological arrangements of Julius
Caesar, to fall on March 25th. This explains the date of
'*Octavo Kal. Ap.'* in the above quotation. After the Equinox had been fixed (as some say by the Council of Nicaea)
on March 21st, ecclesiastical ingenuity was equal to the
occasion in discerning other *a priori* reasons for the fitness
of things on *this* hypothesis. The first day of Creation was
then found to fall most appropriately on March 20th, when
it was pointed out God having created light, ' divided the

light from the darkness,' *i.e.* made the light and darkness equal, which clearly points to the Equinox immediately following !

IV. *Note on Mazzoni's argument in his 'Difesa di Dante.'*

[In reference to p. 24.]

Mazzoni's discussion of *Inf.* xx. 127 in conjunction with *Purg.* ix. 1, &c. in his *Difesa di Dante*, I. c. 76, is very curious. He seems to regard it as axiomatic in regard to *Purg.* ix. 1, &c. (1) That the Lunar Aurora is referred to ; (2) that lines 7, 8 describe $2\frac{1}{2}$ hours of night or a little later : and (3) as regards *Purg.* xviii. 76, that the hour of moonrise is referred to. He quotes two passages of Pliny respecting the retardation in the rising of the Moon, and on the strength of these, constructs two elaborate Tables for finding the time of moonrise after sunset from Full to New. Next, describing the night referred to in *Purg.* ix. as the *third*, and that in xviii. 76 as the *fourth* day of the Moon's age, he finds Dante *prima facie* quite wrong in both cases, since in the former he implies a retardation of more than $2\frac{1}{2}$ hours instead of $1\frac{7}{12}$ hours, as indicated by Pliny's Tables (p. 306) : and in the latter he implies at least 4 hours, whereas the correct time would be $2\frac{9}{34}$ hours (*ib.*). Mazzoni proceeds to defend him by explaining that Full Moon was on Monday, April 4th, in 1300. On Tuesday evening Dante entered the Inferno, and on the night of the 6th escaped from the perils of the Inferno 'per dar principio ad un altro viaggio.' (Here follows a very curious passage to prove that this day was specially chosen because on that day, *viz.* April 6th, were done by the Ancients many valorous actions, according to Aelian, whom he quotes to show that on that day were fought the Battles of Marathon,

Plataea, and Mycale : also Darius was defeated by Alexander
the Great; and moreover it was the birthday of Socrates!
Hence Mazzoni thinks the day was significantly chosen by
Dante for this part of his journey : and that in this he was
wiser and more pious than Petrarch, who chose the same
day for the commencement of his love[1].) He thinks it
reasonable that a day was spent in the *ascent* from the
centre of the Earth to its surface at the Mountain of
Purgatory, since the same amount of time was spent in
the *descent* of the Inferno. This he justifies in a quaint
matter-of-fact manner thus : ' compensando la malagevo-
lezza della salita, colla tardanza che s' era fatta nella scesa
per ragionare con molte anime ! ' He is able thus to argue
that the passage in *Purg.* ix. 41, &c. may relate to the fourth
night there (though only the third night in our hemisphere),
when, according to Pliny's computation, the moon would
rise at $2\frac{27}{48}$ hours of night. He rather lamely continues that
as on the next night, according to Pliny's rule, the moon
would rise at $3\frac{20}{48}$ hours of night, that might be described
as about 4 hours, and therefore ' quasi a mezza notte (?).'
He seems to feel, however, that his laboured ' difesa ' of
Dante has not been entirely successful, as he adds that
Dante probably took account of the ' velocità del moto,
ch' ella aveva in quel tempo, per partirsi dalla oppositione . . .
seconda la quale non ci ha Plinio lasciata regola alcuna ! '
(p. 309.)

V. *On the position of the Earthly Paradise and Purgatory.*

[Note on p. 61.]

In reference to the Earthly Paradise we may note that
the curious notion of the Euphrates and Tigris having a

---

[1] Mazzoni doubtless refers to the passages already cited, *sup.* p. 20.

common source (see *Purg.* xxxiii. 112) is frequently found in other writers, including at least one profane, as well as many Christian authors. In the case of the latter, and others acquainted with the Vulgate, it is no doubt derived from *Gen.* ii. 10 and 14, where a ' fluvius' rising in Paradise parts into four streams, two of which are Tigris (in Engl. Version 'Hiddekel') and Euphrates. It is not so easy to account for its appearance in such an author as Sallust (*Hist. Frag.* cited by Isid. *Orig.* xiii. c. 21 § 10) : 'Sallustius, auctor certissimus, ita asserit Tigrim et Euphratem uno fonte manare in Armenia.' Boethius (aptly quoted by *Scart.*) in *De Cons.* Lib. v. Metr. 1, writes :

> Tigris et Euphrates uno se fonte resolvunt
> Et mox abjunctis dissociantur aquis.

Dante may have derived the idea either from Boethius or Isidore or perhaps from his own master Brunetto Latini who writes thus in *Tes.* iii. c. 2 : 'Salustio dice che Tigris et Euphrates che passono per Armenia escono d' una medesima fontana.' Brunetto also says of the Euphrates ' corre per Armenia et movesi dal paradiso terreno.' S. Thomas Aquinas notes this geographical difficulty in Genesis, and states it as one of the arguments against Paradise being a ' locus corporeus.' He replies to it by adopting the solution of S. Augustine, who had previously observed the difficulty, and deals with it thus :—' Ea flumina quorum fontes noti esse dicuntur alicubi esse sub terras, et post tractus prolixarum regionum locis aliis erupisse, ubi tanquam in suis fontibus nota esse perhibent' (*De Gen. ad lit.* Lib. viii. pp. 612-3, Ed. Basil. 1556). He proceeds further to identify Geon or Gihon with the Nile, and Phison with the Ganges. This identification of the four rivers however is not first found in S. Augustine. It occurs for instance in

Josephus, *Ant. Jud.* I. iii. § 3; and many other writers are cited as adopting it by Corn. a Lap. in his Commentary on Genesis *h. l.* It is commonly repeated in later times, as by Isidore (*Orig.* xiii. 21); Sir John Maundeville (with much quaintness, see pp. 304–5, Ed. 1886); B. Latini (at any rate as regards the Nile, *Tes.* iii. c. 2); and it is also found in the Mediaeval Mappae Mundi, *passim.* Rabanus Maurus, *In Gen.* Lib. I. c. xii. reproduces the explanation and almost the words of S. Augustine. So also does Gervase of Tilbury, *Ot. Imp.* i. 11, &c., &c.

As regards the situation of Purgatory the various opinions, as well as the general consensus, on this subject may be gathered from the exhaustive discussion by Bellarmine, *De Purgatorio,* Lib. II. cap. vi. headed '*De loco Purgatorii.*' He there collects and criticises all the views known to him as having been held on this subject, to the number of eight, but there is no trace in them of any notion that Purgatory was in the Southern Hemisphere, or indeed on the earth's surface at all. He then gives the following opinion as 'communis Scholasticorum, Purgatorium esse intra viscera terrae, inferno ipsi vicinum. · Constituunt enim scholastici communi consensu intra terram quattuor sinus, sive unum in quattuor partes divisum, unum pro damnatis, alterum pro purgandis, tertium pro infantibus sine Baptismo obeuntibus, quartum pro justis qui moriebantur ante Christi passionem, qui nunc vacuus remanet.' (This last statement may be illustrated by *Inf.* iv. 52–63 and xii. 38–9.) So again the various legends collected by Wright in his '*Purgatory of St. Patrick*' generally agree in supposing it to be underground, various opinions prevailing as to the spot in the Earth's surface from which it might be approached. Sometimes, though rarely, it was believed to be in the air. Though it scarcely bears upon this, I cannot refrain from reproducing a quotation given by Wright from a pious Italian writer of the 17th century, as showing how very seriously literal was the belief in the topography of these nether worlds. After stating that it was most certain and beyond all doubt that Hell was situated in the centre of the earth, this writer gravely answers the objection that the space would be insufficient for the constantly increasing multitude of the lost, by stating that the souls of the damned could not expect to be allowed so much room as the blessed spirits in Paradise ! As Bartoli says, ' L' Inferno diventa nel concetto medievale un capitolo di geografia.' He mentions that the island of S. Brandan's vision was marked in maps, that it was once formally ceded by Portugal to Castile, and that even in 1721 a Spanish expedition was fitted out to discover and explore it !

It may be worth while to draw attention here to another point for which (as far as I know) Dante had no patristic or scholastic authority, *viz.* the introduction of a frozen region in Hell. The idea however is found in some of the current mediaeval visions which preceded Dante ; *e.g.* in the vision of Tundalus (an Irish monk), which is placed in A.D. 1149, it occurs in a connexion something like that in which it is found in Dante, *viz.* he represents that a winged monster (not however described as Lucifer) is seated in ice devouring sinners who are also embedded in it. ' Sedebat autem haec bestia super stagnum glacie condensum et devorabat animas' (*De Raptu Animae Tundali*, c. vii). In this case this punishment is reserved for licentious Monks and Ecclesiastics. So again in the very singular 11th century *Visione di San Paolo* (according to Tommaseo, of Anglo-Norman origin), ' gli dimoni si ardevano la metade, e l' altra metade affreddavano.' (Scripture of course only describes S. Paul as having had a vision of Paradise—see 2 Cor. xii. 1–4—but mediaeval fancy extended this to the Inferno also, and it is interesting to note that Dante appears distinctly to adopt this legend in *Inf.* ii. 28, &c.). In the still more celebrated *Visione di Frate Alberico* (to which some have maintained that Dante was largely indebted) the lascivious are imbedded in ice to various depths in proportion to their guilt, like Dante's tyrants in the river of blood (*Inf.* xii. 121, &c.). Cold is described as one of the torments of *Purgatory* (though not apparently of Hell) in the 'Visio Drycthelmi' narrated by Bede (*Hist. Eccl.* Lib. V. c. xii), and it appears also in the *Purgatory of S. Patrick.*

VI. *On alleged notes of Time in the Paradise.*

The passage in *Par.* xxii. 151, &c. is sometimes quoted as a proof that 16 hours had elapsed to this point, *i.e.* starting from the very doubtful assumption that Paradise is entered at noon of Purgatory (or midnight of Jerusalem) on the Wednesday. (See on this point the notes on pp. 10 and 54.) If this be so, the indication is a very obscure and inferential one, and quite unlike those that we have met with in the other two Cantiche. I do not feel at all clear that it is so intended, and regard it as a purely ideal astronomical description, and so far as it has a literal interpretation, it is local rather than temporal. In any case the temporal details apply (if at all) only to this earth (*aiuola*), and not to the place where Dante was. I said the whole situation is a purely ideal one, because as soon as we attempt to draw definite inferences here (or in the somewhat similar passage in xxvii. 79, &c.) as we have done in regard to the precise indications given in the *Inferno* and *Purgatorio*, we begin to discover that we are in a sphere to which the ordinary conditions of time and space no longer apply: *e.g.* lines 151–3 imply that the whole inhabited hemisphere of the world was visible. Now remembering Dante's geographical theories already expounded, and having regard to the clear statements of *Par.* xxvii. 82–87, we observe (1) that such a vision as xxii. 151–3 would imply the condition that *both the spectator himself and the Sun* were on the meridian of Jerusalem : and (2) that such a condition was now impossible, since the spectator was in Gemini, and the Sun well advanced in Aries, 'un segno e più partito' (xxvii. 87). Hence as Della Valle (who however struggles manfully with the difficulties of these two passages) says in reference to xxii. 150, &c., 'Il luogo dove il poeta si trova è al tutto arbitrario' (p. 119): and again (p. 120) he states that the position of Gemini is 'non per legge astronomica ma solo per arbitrio e finzione del poeta, al pari di pianeti' (see l. 144, &c.). We might indeed justify this in Dante's own words—

> Chè dove Dio senza mezzo governa
> La legge natural nulla rilieva.
>
> (*Par.* xxx. 122–3.)

I do not consider therefore that the discussion of these passages falls within the scope of our present subject. At the same time I admit that (as I have already said) Dante intends to give us generally to understand that though himself beyond the limits and conditions of time, still the time passing meanwhile on this earth was such that when he returned to it after his ecstatic vision of Paradise, it would be found to be the evening of Thursday, April 14th. (See *sup.* p. 59.)

VII. *Note on Della Valle's '* Senso Geografico-Astronomico *dei luoghi della Divina Commedia.'*

I tried in vain to procure this work (which is out of print) before writing the above, and only succeeded in meeting with it after I had nearly finished, so that when Della Valle is cited, it is generally on the authority of Scartazzini. His conclusions agree in a general way with my own, as may be expected from the fact that he states among his fundamental data that the Paschal Full Moon was on April 7th, and does not appear aware of (or at least makes no allusion in the body of his work as far as I can find to) any other view[1]: *i.e.* the disputed point as to the Calendar and Real Moon is not referred to. I find too that to a great extent we go over the same ground, but my work is mainly independent of his, except so far as sometimes Scartazzini's notes convey Della Valle's results.

In discussing *Inf.* xx. 126 and xxix. 10 (pp. 15 and 21) his conclusions at first are the same as mine as to the hours indicated, but these are corrected later on in a ' Retractation ' (p. 68), (when he is discussing *Purg.* xviii. 76), by the addition of about one hour. This is to suit a theory then started, that the Moon was Full shortly after midnight on *Wednesday April 6th,* which would be, ' secondo il computo e la regola della Chiesa ' a part of April 7 (p. 66). This theory is due to the supposed necessity of bringing moon-*rise* (for so he interprets the passage) nearer to midnight on the evening referred to in *Purg.* xviii. 76. I observe too that he calls that evening *Tuesday,* but as I have argued (see p. 55, etc.), it must surely be taken to

---

[1] The same seems to be the case with Pasquini, who is satisfied with referring to *L'art de vérifier les Dates,* ' opera . . . da fidarsene ad occh chiusi ' (p. 253).

be *Monday.* The result is that he gains (as compared with my calculation) about two days. In other words, his result is just the same as if for that passage he had adopted the *Real* instead of the *Calendar* Moon. In reference to what I have said about the difficulty resulting from this supposition in *Purg.* ix. 1–9, this does not affect Della Valle, since he explains that passage of the *Solar* Aurora of the *Northern* Hemisphere in l. 1, in contrast with the evening hour 'nel loco ov' eravamo' in l. 8. This is of course open to the obvious objection that the glowing and vivid description of verses 1–6 refer to an absent and invisible scene, besides others which I have noticed in the discussion of that passage.

It should be added that near the end of the *Supplemento* to his work Della Valle for the first time (I believe) refers to the difference between the Ecclesiastical (or Calendar) and Astronomical Moons. He makes the latter, however, fall by his calculations on *the tenth of April* in that year, and then proceeds to enunciate the principle for which I have contended, 'che Dante seguiva le opinioni correnti di allora, e sopratutto stava colla regola della Chiesa' (p. 43). How far Commentators generally have been from recognizing this principle we have already seen; and further how Della Valle himself practically departs from it by putting a non-natural sense on *Inf.* xx. 126, and supposing the Full Moon to have been not '*iernotte*' as that term would be ordinarily understood, but on the previous night. This necessitates his adoption (*inter alia*) of the almost demonstrably false interpretation of *Inf.* xxi. 112 as being 10 a.m. instead of 7 a.m. This hour of 10 a.m. he simply assumes without (as far as I can see) a word of justification, and alludes to it more than once as if no doubt existed on the point.

K

VIII. *Note on the Commentary of Talice di Ricaldone.*

Since these sheets were in the Press I have obtained a sight of the hitherto unpublished Commentary on Dante by Talice di Ricaldone, which has just been printed in a magnificent folio by order of the King of Italy. We have here the notes apparently of some Lectures written or delivered in 1474 *in burgo Liagniaci* (*i.e.* probably Lagnasco, near Saluzzo. See *Pref.*). He scarcely touches on any of the points discussed in this Essay, and passes over in complete silence the chief passages in which difficulties of date or time are involved. In his Note on *Inf.* i. 1 however he declares distinctly for the year 1300.

I note the following brief extracts as having some bearing on points we have discussed :—

On *Inf.* xxi. 112 he refers to the different ways of counting years A.D. in the words : 'Vel describit tempus more Tuscorum qui describunt annos ab incarnatione, et nos a nativitate.' He also declares, like the other early Commentators, for the hour of 7 and not 10 a.m., though by a clerical error he writes 5, thinking no doubt of the word *cinque* in the text. His words are :—'Deus passus est hora sexta, et erat una hora diei ita quod 5 hora' (*sic*).

On *Purg.* ix. 1, &c. he comments thus :—'*La concubina*, &c., id est aurora lune. . . . Et intelligunt autores de aurora solis. Sed Dante intelligit hic de aurora lune, et in hoc facit novam fictionem, ita quod aurora solis est uxor Tithoni et aurora lune est amica Tithoni. . . . *Del freddo animale*, ab effectu. Illud signum effective est frigidum, et est tristius signum quod sit in caelo. Unde illi qui orti sunt sub scorpione semper faciunt vilia officia'!

He takes *Purg.* xviii. 76 apparently as not referring to moonrise, since he comments thus :—'Et describit tempus dicens quod luna clare lucebat ita quod offuscabat multas stellas splendore suo : et erat quasi media nox.'

IX. *Note on Vedovati's Esercitazioni Cronologiche, etc.*

[See p. 5.]

This work also has only come into my hands since I began to print. I therefore add a few notes here. Vedovati maintains that though the Florentine usage was to count years *ab Incarnatione,* Dante would certainly have counted them, as usually, *a Nativitate* (see *sup.* pp. 48, &c. and Vedovati, pp. 15, 23, &c.). The reason for this assertion does not seem clear. Still less do I understand the statement made later by Vedovati (p. 27): 'L' anno 1300 *ab Incarnatione Verbi Divini,* in ultima, corresponde di fatto al 1301 *a Nativitate Domini.'* (Is not the reverse the case ?) In regard to the prophecies of *Inf.* vi. 68 and x. 79 (referred to in a previous note), Vedovati strongly maintains that 'tre *soli'* are three *days, viz.* Nov. 2, 3, and 4 in 1301, after the entrance of Charles of Valois into Florence ; and he advocates the strange notion that in the latter passage 'cinquanta volte' means *Quarters* of the Moon (!), so that the period would be rather less than a year, and he becomes quite jubilant over this undoubtedly original suggestion (see p. 21).

Further he states (I know not on what imaginable grounds) that the Paschal Full Moon of 1300 was on Palm Sunday, April 3 [1] ! (p. 27), and he adds : 'E perciò la Luna del Giovedì Santo di sera non poteva esser *tonda,* come più volte afferma Dante.' By the way, he prudently omits to show how these statements of Dante can be made to square with his own date of 1301, which they certainly do not. (See the Calendar of 1301.)

[1] So also, as we have seen, Ponta (*sup.* p. 27).

K 2

Though the point does not bear on our subject, I may perhaps mention another paradoxical theory of Vedovati, *viz.* that Benedict XI is the *Veltro* of *Inf.* i. 101. As Benedict died in 1304, probably before the passage was written, and certainly long before it was 'published,' it is inconceivable that Dante should have given a permanent and very prominent position to a prophecy which had so obviously failed of its accomplishment.

X. *On the unity and symmetry of the plan of the Purgatorio.*

[As a note in explanation of Table No. VII.]

Lest any surprise should be felt at the minutely planned and connected scheme of the whole Purgatorio, implied by the series of time-references which I have traced out, it may be worth while to draw attention to the still more striking proof that through the whole *Cantica* ' one unceasing purpose runs,' which is afforded by the annexed Table (No. VI).

Note here the following points of similarity throughout the description of the Seven *Cornici* :—

1. At the *commencement* of each, the Poets are greeted with examples of the *Virtue* which is opposed to the Vice being expiated in that *Cornice*, that Vice being in each case one of the Seven Deadly Sins recognized by the Church.

2. Towards the *end* of each, there are similar examples of each *Vice* held up to odium.

3. In each case the examples are taken alternately from Sacred and Profane history, except in the fifth *Cornice*, where the examples of the Vice alternate thus *in groups.* This systematic balancing of Sacred and Profane illustrations is found also in the *Paradiso* several times, *e.g.* iv.

83–4 (S. Laurence and Mucius); v. 66–70 (Jephthah and Agamemnon); viii. 130–2 (Esau and Jacob and Quirinus)[1].

4. There is an obvious correspondence of the *number of examples* given of each Virtue and its corresponding Vice, though sometimes one or even two Cantos intervene between them. In *Cornici* i. and v, three *instances* in the one case are balanced by three *groups of instances* in the other. In the former *Cornice*, the groups are marked off from one another by the repetition in twelve Successive Terzine, of the initial words, *Vedeva, O,* and *Mostrava,* which are again repeated and gathered up into the three lines of a concluding Terzina (see Canto xii. lines 25–63)[2]. In the latter, the groups are marked off by an arrangement peculiar to that *Cornice, viz.* putting *together* two or more instances from Profane and Sacred History respectively instead of making the instances alternate. The only slight exception to this symmetry seems to be in *Cornice* vi, where the number of examples of the Virtue and of the Vice do not correspond.

5. Note especially that in *every* case some incident in the life of the Blessed Virgin is the first instance of the Virtue held up for admiration.

6. In every case also they are dismissed from the *Cornice* with the utterance by an Angel of a portion of one of the Beatitudes from the Sermon on the Mount. (In the 1st *Cornice* (xii. 110) it is true the Angel is not expressly mentioned, and there is some doubt as to the reference of *voci* in l. 110. See Scartazzini's note, *h. l.*)

---

[1] So in the *Ep.* to Can Grande, § 1, an illustration is drawn from the Queen of the South and Pallas.

[2] There is a very similar case of a Canto symmetrically constructed in *Par.* xix. 115–141. Compare also the six times repeated *Ora conosce* at the commencement of alternate *terzine* in *Par.* xx. 40–70.

This last is a peculiarly interesting point, as it establishes I think beyond all possibility of doubt the reading *Sitiunt* instead of the commonly received *Sitio* in *Purg.* xxii. 4. This is the more interesting from the fact that the reading seems to have disappeared almost entirely from our existing MSS. In fact, it may be said to be practically devoid of MS. support. I have only met with it *once*, though I have searched for it in more than 180 MSS. This was in a MS. in the Corsini Library at Rome (No. 346 in *Colomb de Batines*). I find it registered, however, as occurring in two MSS. of some celebrity at Udine, known as the *Codice Florio* and the *Codice Bartoliniano*, which are stated by Colomb de Batines to have closely related texts.

Other instances of this unity of design may be found in the way in which the first three *Cornici* are assigned to the first day and the last three to the second day, while the central *Cornice* (in which *Accidia* is purged) appropriately occupies the intermediate *night* (see *sup*. p. 106). Also in the fact that a mystical dream occurs each day in the hour before sunrise (for this also see *sup*. p. 98).

LIST OF TABLES.

TABLE I.

Calendar, 1300.

MARCH.

21	M	SPRING EQUINOX.
22	T	NEW MOON (Real).
23	W	
24	Th	NEW MOON (Calendar).
25	F	LADY DAY.
26	S	
27	**S**	PASSION SUNDAY.
28	M	
29	T	
30	W	
31	Th	

APRIL.

1	F	
2	S	
3	**S**	PALM SUNDAY.
4	M	
5	T	FULL MOON (Real, 3 A.M.)
6	W	
7	Th	FULL MOON (Calendar).
8	F	GOOD FRIDAY.
9	S	EASTER EVE.
10	**S**	**EASTER DAY.**
11	M	
12	T	
13	W	
14	Th	

TABLE II.

Calendar, 1301.

MARCH.

24	F	FULL MOON (Real).
25	S	LADY DAY.
26	**S**	PALM SUNDAY.
27	M	FULL MOON (Calendar).
28	T	
29	W	
30	Th	
31	F	GOOD FRIDAY.

APRIL.

1	S	EASTER EVE.
2	**S**	**EASTER DAY.**
3	M	
4	T	
5	W	
6	Th	

TABLE III.

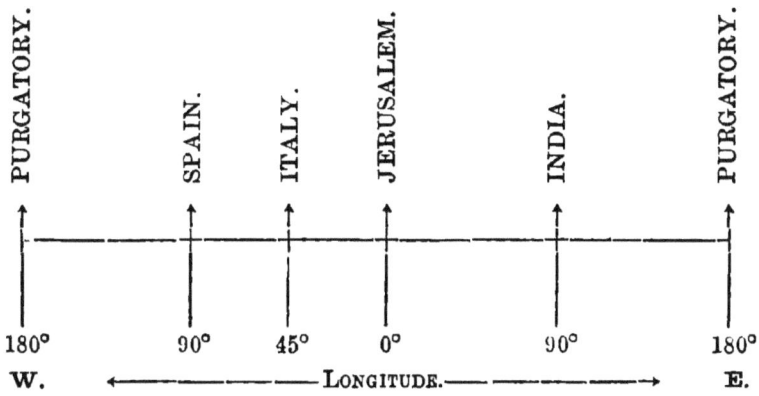

TABLE IV.

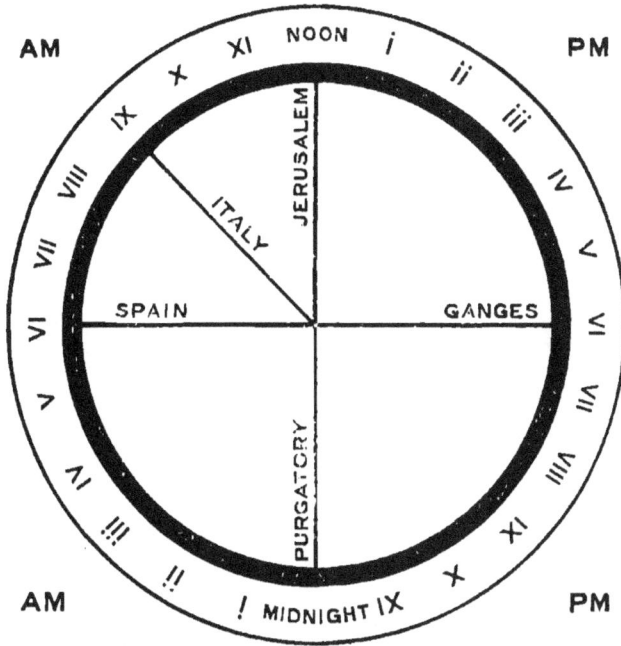

Directions.—If the part of this diagram within the dark circle be cut out separately in cardboard, so that it can be made to revolve, it will be possible to see at a glance the simultaneous hours described in the five passages of the *Purgatorio* discussed on pp. 70–73.

	SELVA OSCURA	Th.	All night.	i. 21.
	TRE FIERE, &c.	Fr.	All day.	i. 37, &c.
CIRCLE I.		–	Nightfall.	ii. 1.
— II.		–		
— III.		–		
— IV.		–	Midnight.	vii. 98.
— V.		–		
— VI.		Sat.	4 A. M.	xi. 113.
— VII.	1st GIRONE			
	2nd			
	3rd			
— VIII.	1. BOLGIA			
	2.			
	3.			
	4.		6 A. M.	xx. 125.
	5.		7 A. M.	xxi. 112.
	6.			
	7.			
	8.			
	9.		1 P. M.	xxix. 10.
	10.			
— IX.	CAINA			
	ANTENORA			
	TOLOMEA			
	GIUDECCA		7.30 P. M.	xxxiv. 96.

TABLE VI.

EASTER-DAY,		1. 19–21	c. 4 A. M.
		– 107–115	c. 5 A. M.
Ap. 10.		2. 1	c. 5.15 A. M., sunrise.
	ANTE-	– 55–7	6 A. M.
		3. 16, 25	6 to 6.30 A. M.
		4. 15	c. 9 A. M.
	PURGATORY.	– 138	Noon.
		7. 43	Day declining.
		– 85	'Poco sole.'
		8. 1	Just after sunset.
		– 49	c. 7.30 P. M.
		9. 1–9	c. 8.45 P. M.
MONDAY,	PURGATORY.		
		– 13, 52	Before dawn.
11.		– 44	c. 7.30 A. M.
	CORNICE. I. {	10. 14	c. 8.30 A. M.
		12. 81	c. Noon.
	— II.	15. 1	3 P. M.
	— III. {	– 141	c. 6 P. M.
		17. 9	c. 6.30 P. M.
	— IV. {	– 62, 72	Twilight.
		18. 76	Towards midnight.
TUESDAY,	— V. {	19. 1–6	c. 4.30 A. M.
		– 37	Full daylight.
12.	— VI. {	22. 118	11 A. M.
		25. 1–3	c. 2 P. M.
		26. 4–6	c. 4 or 5 P. M.
	— VII. {	27. 1–5	c. 6 P. M.
		– 61	Sunset.
		– 70	Twilight.
		– 89	Starlight.
WEDNESDAY,	EARTHLY	– 94	Before dawn.
13.		– 109, &c.	Sunrise.
	PARADISE.	– 133	Sun fully up.
		33. 103	Noon.

Table VII.

CORNICE →	1. PRIDE.	2. ENVY.	3. ANGER.	4. 'ACCIDIA.'	5. AVARICE.	6. GLUTTONY.	7. LUST.
Ex. of VIRTUE.	x. 40. 1. B. M. V. 2. David. 3. Trajan.	xiii. 28. 1. B. M. V. 2. Orestes.	xv. 88. 1. B. M. V. 2. Pisistratus. 3. Stephen.	xviii. 100. 1. B. M. V. 2. Julius Caesar.	xx. 19. 1. B. M. V. 2. Fabricius. 3. *S. Nicholas*	xxii. 142. 1. B. M. V. 2. Ancient Roman Women. 3. Daniel. 4. The Golden Age. 5. S. John Baptist.	xxv. 128. 1. B. M. V. 2. Diana.
Ex. of VICE.	xii. 25. 1. { Lucifer. Briareus, &c. Nimrod. 2. { Niobe. Saul. Arachne. Rehoboam. 3. { Eriphyle. Sennacherib. Cyrus. Holofernes.	xiv. 133. 1. Cain. 2. Aglauros.	xvii. 19. 1. Procne. 2. Haman. 3. Amata.	xviii. 133. 1. Israelites in Wilderness. 2. Companions of Aeneas.	xx. 103. 1. { Pygmalion. Midas. 2. { Achan. Ananias and Sapphira. Heliodorus. 3. { Polymnestor. Crassus.	xxiv. 123. 1. Centaurs. 2. Companions of Gideon.	xxvi. 40. 1. Sodom, &c. 2. Pasiphae.
BEATITUDE.	xii. 110. B. pauperes spiritu.	xv. 38. B. misericordes.	xvii. 68. B. pacifici.	xix. 50. B. qui lugent.	xxii. 5. B. qui sitiunt.	xxiv. 151-4. B. qui esuriunt.	xxvii. 8. B. mundo corde.